First Steps with Puppies and Kittens

First Steps with Puppies and Kittens

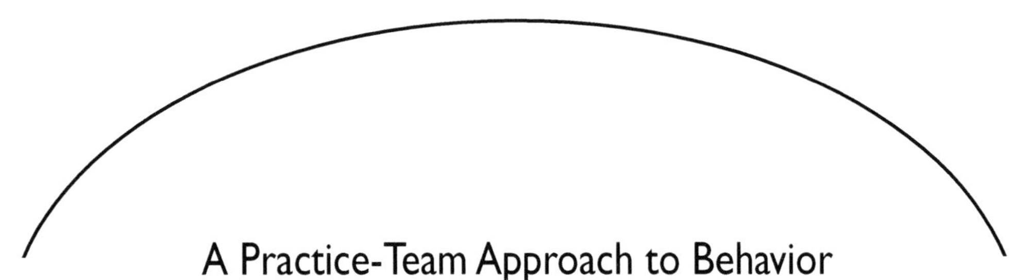

A Practice-Team Approach to Behavior

Linda M. White

Reviewed by Lisa Radosta, DVM, DACVB

With contributions by Christine Merle, DVM, MBA, CVPM

With contributions by John White

AMERICAN ANIMAL HOSPITAL ASSOCIATION

press

American Animal Hospital Association Press
12575 West Bayaud Avenue
Lakewood, Colorado 80228
800/252-2242 or 303/986-2800
AAHAPress@aahanet.org
www.aahanet.org

ISBN 978-1-58326-101-9

Library of Congress Cataloging-in-Publication Data to come

Design by Elizabeth Lahey.

Cover photographs courtesy of in order of appearance top to bottom, left to right, Aleris Wang, Herve de Brabandere, Elaine Tan, Patrizia Schiozzi, Bethan Hazell, and Alexis Lecomte.

DEDICATION

This book is dedicated to all of the wonderful animals I have had the pleasure of working with and to our girls who have shared and are sharing their lives with us...

Typsi, the first dog I ever trained, taught me how important patience is when working with animals and how forgiving animals can be to our shortcomings.

Taffy taught me to never give up on anything that is important to me and has made me a believer that animals are psychic!

Bunny taught me it is okay to walk off the beaten path in life and that life was meant to be enjoyed and not taken so seriously. Her self-confidence, poise, and warm, outgoing personality were always a joy in our lives. She believed, as all animals do, to simply be in-the-moment.

Gabrielle's lessons are more subtle. She is teaching me the art of gentle persuasion and that if you focus and are patient, you will get what you want in life. She is a caring and serious soul, willing to lie beside the ill to offer comfort. She also is athletic and a tomboy when the opportunity presents itself. She has shown me how to take a multifaceted approach to life.

TABLE OF CONTENTS

CHAPTER 3:
PATIENT BEHAVIOR ADVOCATE PET TRAINING PRINCIPLES   27

CHAPTER 4:
PUPPY BEHAVIOR SESSIONS 51

CHAPTER 5:
KITTEN BEHAVIOR SESSIONS 107

APPENDIX A:
PUPPY QUESTIONNAIRES AND CHECKLIST

APPENDIX B:
KITTEN QUESTIONNAIRES AND CHECKLIST

APPENDIX C:
SAMPLE CLIENT HANDOUTS (TRAINING PRINCIPLES)

ABOUT THE AUTHOR

Linda White grew up on a farm in Mosinee, Wisconsin, where her love affair with and understanding of all animals began at a young age. She currently resides in Phoenix, Arizona, with her husband John, her long-haired dachshund Taffy, and her standard poodle Gabrielle.

Linda is an internationally recognized author, speaker, and professional animal trainer. She has more than 30 years of hands-on experience helping families and veterinarians find solutions to pet behavior challenges. Her emphasis is on puppy and kitten development and training. She is passionate about helping pet parents become successful trainers and preventing behavior challenges from becoming behavior problems.

Linda is president of PuppySmarts®, which publishes her award-winning video training lessons that teach pet parents how to successfully train their puppies. She is also president of Linda White and Company, which focuses on helping veterinarians help pets. She writes books and articles, gives speaking presentations, and offers consulting services.

Linda is an active member of VetPartners, the Association of Professional Dog Trainers, and the National Speakers Association. Linda can be reached for questions and comments at Lindaw@FirstStepswithPets.com.

ACKNOWLEDGMENTS

I have been blessed with guardian angels here on earth. They have helped with this book in many ways. First and foremost is my husband, John, who has given so much time, patience, support, and understanding over the many months it took to gather and assemble this information. John is my best friend and confidant, and I love him dearly.

I give thanks for my son Kevin, whom I am so grateful to have in my life. He offers this world his own special love and understanding of both humans and animals, much like his grandfather did. His ability to understand others and willingness to help is nothing less than amazing. Kevin's lightheartedness and ability to play with life is inspiring.

A special thank you goes to my father, now in heaven, for it was his passion for animals and people that I believe I inherited. He often brought home lost or stray dogs. What was so amazing about this was that he found new homes for all of them as well. He had a special understanding for animals as well as humans who were in need of help or hope. Rich or poor, two-legged or four, we could all count on him to be there when we needed him most.

To my sister Isabelle, my mother Marie, and stepfather Herb, who have encouraged and believed in my life's passion, I offer a grateful thank you, and a special mention to my stepfather Herb for his help in editing this book.

My niece Cassandra did an excellent job of editing some of this material while bringing her perspective to the subject matter. I loved working with Cassie.

I am very fortunate to have such a special family who believes in the work I do and is so willing to help me. I would also like to acknowledge and thank my extended family—the friends I chose as my "family."

The people listed here sincerely care about the veterinary industry and the animals it serves. They are caring, intelligent people and I am grateful to have them in my life. I am humbled by their kindness and their willingness to share their wisdom, knowledge, and talents with me. I thank each of them from the bottom of my heart.

Carol Pernsteiner, DBA, has offered guidance and support from the beginning. Her compassion for animals, and her generosity and giving spirit, helped make this book and the PuppySmarts training lessons a reality. Carol is kind, intelligent, understanding, and a very special person I am lucky and honored to call my dear friend.

Scott Line, DVM, PhD, DACVB, was the first professional in the veterinary industry to endorse the work and the training methods I believe in and teach. In addition, he helped with the editing of some of the forms within this book. Scott is a gentle and kind soul.

Elizabeth Bellavance, DVM, MBA, CMA, gave a great amount of time and energy to the work I was doing. She also helped by editing some of the forms in the appendices.

Thomas Catanzaro, DVM, MHA, FACHE, is a consultant who worked quite hard at trying to help me understand the training void in the industry and how I could fill it. He is probably the most forward-thinking person I will ever meet. His book,

Promoting the Human-Animal Bond in Veterinary Practice, gave me a clearer understanding of what a true client-based practice could look like

Kirk Augustine, president of Foray's Inc., and an industry consultant, helped me to understand the meaning of the word "focus," how to zero in on what was important to me. He has played an invaluable role in helping me move to the next level of where I wanted to go. His clarity of the steps that need to be taken has always been right on target. I am blessed and honored to call Kirk and his wife Cindy my friends.

Lisa Radosta, DVM, DACVB, reviewed the content of this book to assure its alignment with the teachings of the veterinary behaviorist community. She found the time to devote to this work, even with the arrival of her baby girl, Isabella. Lisa and I met in 2002. She is a passionate veterinarian with a wonderful understanding of animal behavior to which she has dedicated her life's work. I am grateful she was so willing to help with this book as well as for her friendship, openness, and integrity.

Lorna Zwerin-Parson, MSHR, SPHR, helped with the final edit of this book. Lorna and I have never met in person, yet this woman was willing to offer a helping hand. Lorna and I serve together on the membership committee for VetPartners. She is a giving person, and I appreciate her friendship.

Roger Cummings, CVPM, knew where I needed to be six years ago! His marketing skills coupled with his understanding of the veterinary community have helped bring additional clarity to my vision and goals. Roger is a wonderful gentleman and friend.

INTRODUCTION

Thank you for your commitment to the overall well-being of the patients you serve.

This book offers a practical hands-on approach for veterinarians and team members to address puppy and kitten behavioral needs before they become behavior problems. Behavior information in this book is a team function under veterinarian guidance. This book captures the author's experience in training puppies, kittens, and clients to be successful together. This book is therefore different from and a complement to the good theoretical books available that provide a deeper understanding of the "how" and "why" behind behaviors.

Clients need a simple, easy-to-understand approach for help in integrating their puppies or kittens into their families. Veterinarians and their team members are the professionals that the majority of puppy and kitten parents will turn to for help when settling their new pets into their homes. As advocates for patients, you are their best hope for success.

While observing veterinary practices over the past several years, I became acutely aware of how busy most practices are. The time devoted to new puppy/kitten appointments ranges from 30 to 120 minutes, covering zoonotic disease, care and feeding, grooming, behavior, spay/neuter, and vaccinations, and the list goes on. Clients experience a tremendous information overload and they feel overwhelmed. As a result, many recommendations are forgotten. Practices know that offering patient training and behavior programs is essential, but finding the time to implement such programs can seem next to impossible.

The approach outlined in this book is a road map for successful implementation. It uses a practical approach to show, step-by-step, how you can find the time and make the greatest contribution to puppies, kittens, and clients. It is important before you move from one step to the next that the entire team is on board, comfortable, and ready to move forward. You can customize the suggested standards of care, protocols, questionnaires, checklists, and programs within this book to fit the practice's needs. Different forms can be printed out as needed from the companion CD.

Most people enter this industry because they care passionately about animals. I am enthusiastic about helping clients successfully settle their newest family members into their homes. Combining our passions, we will keep healthy, happy pets with their new families. This book will put the practice in a position that will strengthen and support the relationships between clients and their pets as well as between clients and the veterinary practice.

Chapters 1 and 2 address the managers and owners of the practice. Chapters 3 through 5 address the Patient Behavior Advocates. The forms and handouts in the appendices are contained on the companion CD so they can be modified and used in the practice. The book includes extensive cross-referencing; the reader should use the detailed table of contents to find specific pages for these references to other parts of the book. For ease of reading, puppies are referred to as males, kittens as females, and clients as females, although all pronouns are to include both genders. While I mention several products by name, I do so for illustration purposes only as an example of a product type; they should not be taken as an endorsement or recommendation.

CHAPTER 1:
THE BUSINESS SIDE OF BEHAVIOR

with Christine Merle, DVM, MBA, CVPM, and John White

What Is a Client Worth? Calculating Client Value

Consider this scenario: Mr. and Mrs. Jones just relinquished their German shepherd puppy, Max, because his behavior became more than they could manage. Max and Mrs. Jones have been to your practice a few times. Is your practice going to suffer a loss from this event? How do you know? The answer lies in learning the value of a client in your practice. This is a good financial benchmark for the entire health care team to learn, know, and understand.

Calculating client value provides an understanding of the money and time it requires to obtain a new client. Client value can also be used to evaluate new services to determine whether the investment in that service will be rewarding. Greater net profit is generally achieved from retaining clients and patients long term. Clients and patients have value directly related to the frequency they are seen in the practice.

In its simplest form, client value is the average amount the client spends per year times the average number of years the client stays with your practice. To determine the average amount spent per year, multiply the average number of transactions per client and by your average transaction charge.

How long the average client stays with your practice is called client longevity. It is calculated by dividing the total number of active clients by the number of new clients in that same time period.

So how much were Mrs. Jones and Max worth to your practice? Using the averages of AAHA's *Financial and Productivity Pulsepoints*, Fifth Edition, we get the following:

Amount spent per year = 4.9 transactions × $120 each = $588

Client longevity = 4,046 active clients ÷ 750 new clients = 5.4 years

Client value = $588 spent per year × 5.4 years = $3,175.20

So on average, Mrs. Jones and Max should be worth about $3,175 to the practice. What is a client worth in your practice? You can visit www.FirstStepswithPets.com to use a free client value calculator.

Increasing Client Value

To grow a practice, you can either: (1) see more clients or (2) increase the value of the clients you have. Increasing the number of clients you have is possible, but it is a slow process that involves adding veterinarians, staff, facilities, and marketing. It is much easier and faster to increase the value of the clients you already have. This can be accomplished by using any of the following strategies:

- Increase the number of active clients
- Increase the number of client issues that can be handled by properly trained team members instead of veterinarians
- Increase the number of transactions for each client
- Increase the average value of each transaction
- Increase client longevity

Proactively addressing behavior challenges in young pets before challenges become problems allows you to positively affect all of the above factors, each of which is a part of your revenue and profitability. Using the program outlined in *First Steps with Puppies and Kittens* is also a great marketing differentiator for your practice. It will attract new clients and retain existing ones. Behavior counseling can primarily start as a function of the non-veterinary professional team members. The veterinarian should need to spend only short periods of time initially, and then refer the client to the non-veterinary team member(s) who have learned the behavior programs. The veterinarian may potentially meet the client and patient again for the final progress review with the training team member and prescribe the appropriate tools. With the addition of behavior-specific Patient Behavior Sessions (PBSs), an increase in transactions will occur.

The human-animal bond is a very important concept whose value we are only beginning to understand. Humans who view their pets as part of the family will care for them as if they are family members. The following statistics from the American Veterinarian Medical Association's *2007 US Pet Ownership & Demographics Sourcebook* demonstrate the impact of the human-animal bond on a practice's revenue.

Households that viewed their dogs as family members visited the veterinarian an average of 3.0 times in 2006, compared to 2.2 visits for households that considered their dogs to be pets/companions and 1.1 visits for dogs considered as property. Dog-owning households that considered their dogs to be family members spent 1.7 times more on veterinary expenditures in 2006 than those that considered their dogs to be pet/companions and 3.4 times more than those that considered their dogs to be property.

Households that viewed their cats as family members visited the veterinarian an average of 2.0 times in 2006, compared to 1.4 visits for households that considered their cats to be pets/companions and 0.7 visits for cats considered as property. Cat-owning households that considered their cats to be family members spent 1.6 times more on veterinary expenditures in 2006 than those that considered their cats to be pets/companions and 3.3 times more than those that considered their cats to be property.

The more you help families with pet family member behavior issues, the better off your practice will be. When the human-animal bond is strong and households consider their pets to be family members, they will visit three times more often and spend three times more than households that view their pets as just a "dog" or "cat" they own. The number one cause for the rupture of the human-animal bond is behavior.

Many veterinarians do not believe they lose many clients. It is human nature to remember the clients and patients who have returned to us or those we have seen recently. It is also human nature to forget those patients we have not seen in a while or saw only once. "Out of sight, out of mind" is a real phenomenon. The importance of the role of behavior to the veterinary practice is evidenced by the following realities:

- Ninety percent of pet owners look for help with behavior issues.[1]
- Seventy-two percent of clients ask their veterinarians about behavior problems.[2]
- Behavior is the fourth largest growth area in veterinary medicine.
- The most frequent reason for canine relinquishment is behavior.[3]
- More pets are relinquished or put to death due to behavior disorders than infectious, neoplastic, and metabolic diseases combined.[4]
- The majority of dogs relinquished have been with a family for less than one year.[5]
- Thirty-seven percent of dogs are surrendered during the first three months of ownership.[6]
- The average veterinary practice was estimated to lose $20,000/year in income (*not* revenue) due to the relinquishment of pets for behavior concerns, resulting in lost services,[7] and this is likely a gross underestimate.[8, 9]
- Veterinarians lose approximately 15 percent of their client base annually due to unresolved behavior problems.[10]
- Studies have shown that investments in modern behavior medicine, including the handling of animals in the exam room, are financially sound compared to labor costs, client goodwill, team member burnout, and overall job satisfaction[11]

Increasingly more attention is being paid to the seriousness of dog bites in the United States. The following facts reveal the sad realities:

- Every 40 seconds, someone in the United States seeks medical attention for a dog bite injury.[12, 13, 14]
- Every year, 4.7 million dog bites are reported.[15]
- Most dog-bite victims are children younger than 13 years of age; males and five- to nine-year-olds are bitten most frequently.[16]
- Each year, 800,000 individuals require medical attention for dog bites, with children being the most likely victims.[17]

- Dog bites are becoming an increasingly important public health problem; the number of reported dog bites increased significantly over a 10-year period (33 percent), whereas the dog population increased only 2 percent.[18]

- The face is the most commonly affected site of dog bites (40 percent of the cases).[19, 20]

- The second most common cause of all injuries in children is dog bites (second to baseball injuries).[21, 22]

- Dog bites account for approximately 25 percent of homeowners' insurance claims.

- Greater than 50 percent of bites took place on the pet owner's property.[23]

- Total annual insurance claims for animal bites are estimated to be $321.6 million, with an average claim of $16,600.[24]

- Certain dog breeds have become legislative targets due to perceived behavior problems.

- Owners have been convicted of murder (involuntary manslaughter) for their dog's actions.[25]

Canine behavior challenges affect your clients and your practice. It is not a question of whether you will address common patient behaviors. It is a matter of when will you recognize the elephant in the living room and begin to implement a behavior program.

First Steps as a Market Differentiator

How do you attract your clients? Some of them come from advertising, but most of us will agree that "word of mouth" is the biggest referral base. So, what makes your practice different from the others? How can you pull more clients into your practice, especially if competition is just down the street? For some practices, focusing on a particular species is one way to differentiate; for others, it may be choosing to highlight the equipment they provide.

The focus of veterinary medicine has shifted from treatment to prevention, and with that shift comes a new emphasis on the overall well-being of the animal, including behavior. Dogs and cats are part of the family unit. The line between good dog or cat and bad dog or cat can be very unclear. How do you make sure that pet owners understand the concepts of proper pet behavior and how their pets can become contributing participants in the family dynamics? Being able to help a pet become a good family member is one way to make your practice different from others in the area. Have you noticed all the activities in which the family pet now participates? It may be watching a child's soccer game, picking up the kids at school, or traveling on vacation. It is the good pet that gets to be a part of it all.

Financial Benefits of Behavior

The greatest financial impact of bringing professional behavior counseling into the practice comes from the retention of the client and patient. Remember: 72 percent of pet owners turn to the veterinarian for help with behavior issues.

Basic Behavior Program

For the basic program, a nominal charge to cover the costs of the materials, DVDs, and brochures is added to the existing puppy and kitten vaccine pricing packages or visits. These materials can also be complimentary, as they reduce the amount of veterinarian time spent providing basic behavior information. Addressing behavior issues will improve client retention by helping to prevent a disruption in the human-animal bond—a disruption that often leads to the heartbreaking surrender of a pet.

Intermediate Behavior Program

In the intermediate program, the initial puppy/kitten visit is restructured slightly to focus on the physical well-being of the patient and referral to new-patient orientations, which are groups of approximately 10 clients at a time. These groups, led by non-veterinary team members, cover topics such as medical emergency credit cards like CareCredit®, microchips, pet insurance, proper care and diet, and grooming suggestions, among other topics. These topics do not and should not involve veterinarian time. Once the practice begins to offer new patient orientations, they can also be used as a marketing tool to bring new clients into the practice. The practice can promote the times that such orientations are offered, and can add a fee for the orientation class. A prepayment requirement will increase the client's likelihood of attending the class.

Advanced Behavior Program

The advanced program goes beyond the intermediate program and sets up one-on-one Patient Behavior Sessions with trained Patient Behavior Advocates, who teach clients how to deal with their pets' behavioral needs.

Business Case Metrics

Time and knowledge are what clients expect to receive and pay for from a veterinary practice. This time and knowledge convey value. Products that help us convey knowledge and improve the pet are a portion of veterinary practice, but it is the time and skill of the professionals that the pet owner pays for.

The benefits to a practice of offering a behavior program to a practice include the following:

- The number of active clients will increase. Fewer clients will need to be replaced, so more marketing efforts can go to growing the practice, not just maintaining it.

- The number of client issues that can be handled by properly trained team members instead of veterinarians increases. Veterinarian time can be spent practicing medicine, thus decreasing the cost of providing routine services.
- The number of transactions and value per transaction for each client will increase through the strengthening of the human-animal bond.
- Average client longevity will increase. As more patients keep their pets, client retention improves.

The following cost, revenue, and return-on-investment (ROI) analysis is given as an example. The example is based upon assumptions portraying an "average" practice. Since there really are no average practices, this scenario is intended solely to help any practice understand the tremendous financial impact that incorporating formal behavior counseling programs can bring to the veterinary business.

The assumptions are for a practice with two full-time-equivalent (FTE) veterinarians, 4.3 non-veterinary team members per FTE DVM, and 1,000 active dogs per FTE DVM. The annual attrition of clients, attraction of new clients and patients, and the assumed average costs for non-veterinary team members and annual revenue of $400 per dog for the practice are all very conservative.

Table 1.1 **Basic Behavior Program. Example: per FTE veterinarian per year**

3 minutes added to all visits [2600 visits × 3 minutes × $.83 = $6,474]	($6,474)
Program Sales Revenue [125 × $10 = $1,250, 50% compliance]	$1,250
Subtotal	($5,224)
ROI: Number of dogs saved annually to break even	2.08

Table 1.2 **Intermediate Behavior Program. Example: per FTE veterinarian per year**

3 minutes added to all visits [2600 visits × 3 minutes × $.83 = $6,474]	($6,474)
10 minutes reduced from initial visits [250 visits × 10 minutes × $.83 = $2,075] (opportunity for additional billings)	$2,075
Program Sales Revenue [125 × $10 = $1,250, 50% compliance]	$1,250
A. Orientations [125 × $25/hr = $3,125, 50% compliance]	$3,125
B. Orientations [125 × $30/hr = $3,750, 50% compliance]	$3,750
C. Orientations [125 × $40/hr = $5,000, 50% compliance]	$5,000
Subtotal (A)	($24)
Subtotal (B)	$601
Subtotal (C)	$1,851
ROI: Positive cash flow immediately. Saving more dogs is an additional benefit.	Immediate

Table 1.3 Advanced Behavior Program. Example: per FTE veterinarian per year

3 minutes added to all visits [2600 visits × 3 minutes × $.83 = $6,474]	($6,474)
10 minutes reduced from initial visits [250 visits × 10 minutes × $.83 = $2,075] (opportunity for additional billings)	$2,075
Program Sales Revenue [125 × $10 = $1,250]	$1,250
A. Orientations [125 × $25/hr = $3,125, 50% compliance]	$3,125
B. Orientations [125 × $30/hr = $3,750, 50% compliance]	$3,750
C. Orientations [125 × $40/hr = $5,000, 50% compliance]	$5,000
Patient Behavior Session cost estimate [125 × $25/hr × .5= $1,562]	($1,562)
Patient Behavior Session revenue estimate [125 × $45 = $5,625]	$5,625
Subtotal (A)	$4,039
Subtotal (B)	$4,664
Subtotal (C)	$5,914
ROI: Positive cash flow immediately. Saving more dogs is an additional benefit.	Immediate

Ask yourself: How many pets will your practice see this year that will not return next year because behavior issues are not dealt with effectively, breaking the heart of the pet owner and too frequently resulting in the death of the pet? No one knows the answer. Keeping a single patient in the practice is worth $2,500, or more than $400 per year. Keeping 10 patients is worth $4,000 annually, or $25,000 over the average retention of the client and patient. The financial impact of keeping 5, 10, or 100 more patients adds up quickly. The investment in behavior, combined with the overwhelming statistics of pets relinquished because behavior is not addressed, more than provide reasons for the veterinary practice to take the time to implement good behavior programs for pet owners.

More than the Dollars

Veterinarians, like most of their health care team members, initially came to veterinary medicine because of their love of pets. And veterinarians *do* love pets. Saving pets' lives by addressing behavior challenges in a professional setting makes sense—both financially and emotionally—for the entire health care team. Pet owners who believe the veterinarians and the veterinarian teams in their practice love their pets as much as they do are loyal. These same pet owners believe their pets are members of their families, and they will turn to the veterinary practice for their pets just as they would turn to a pediatrician for their children.

But first, the veterinarian has to believe in the importance of the pets and understand that behavior counseling is appropriate in a veterinary practice. Good medicine is good business. Medicine involves not only many body systems and organs, but also the pets' and pet owners' emotional and psychological well-being.

Endnotes

1. Tom Catanzaro, *Promoting the Human-Animal Bond in Veterinary Practice* (Ames, IA: Iowa State University Press, 2001).

2. American Veterinarian Medical Association, 2003 Survey.

3. Mo D. Salman, Jennifer Hutchison, Rebecca Ruch-Gallie, Lori Kogan, John C. New, Jr., Phillip H. Kass, and Janet M. Scarlett, "Behavioral Reasons for Relinquishment of Dogs and Cats to 12 Shelters," *Journal of Applied Animal Welfare Science*, 3(2) (2000): 93–106.

4. K. L. Overall, "Evaluation and Management of Behavioral Conditions." In: *Clinical Neurology in Small Animals—Localization, Diagnosis and Treatment*, Braund, K.G.(Ed.) (Ithaca, NY: International Veterinary Information Service, 2001). www.ivis.org, Accessed June 2005.

5. Salman *et al.*, 2000.

6. Ibid.

7. Karen L. Overall, "Resolve Common Behavior Problems by Offering Training Advice to Clients," *DVM Newsmagazine*, June 1, 2002.

8. E. R. Shore and K. Girrens, "Characteristics of Animals Entering an Animal Control or Humane Society Shelter in a Midwestern City," *Journal of Applied Animal Welfare Science*, 4 (2001): 105–116.

9. Salman *et al.*, 2000.

10. Jessica Tremayne, "AAFP Pens Behavior Guides for DVMs, Staff, Clients," *DVM Newsmagazine*, April 1, 2005.

11. Karen Overall, "Are you Encountering Problematic Behavior?" *DVM Newsmagazine*, February 1, 2005.

12. www.cdc.gov, Centers for Disease Control and Prevention, Accessed June 2005.

13. www.dogbitelaw.com, Accessed June 2005.

14. www.iii.org, Insurance Information Institute, "Dog Bite Liability," January 2005.

15. Ibid.

16. www.cdc.gov, Centers for Disease Control and Prevention, Accessed June 2005.

17. Ibid.

18. Ibid.

19. Ibid.

20. http://www.phac-aspc.gc.ca/injury-bles/chirpp/injrep-rapbles/dogbit-eng.php, Canadian Hospital Injury Reporting and Prevention Program, "Injuries Associated with Dog Bites and Dog Attacks," Accessed June 14, 2005.

21. www.cdc.gov, Centers for Disease Control and Prevention, Accessed June 2005.

22. Harold B. Weiss, Deborah I. Friedman, and Jeffrey H. Coben, "Incidence of Dog Bite Injuries Treated in Emergency Departments," *Journal of the American Medical Association*,. 279 (1998): 51–53.

23. www.cdc.gov, Centers for Disease Control and Prevention, Accessed June 2005.

24. Ibid.

25. http://www.dogbitelaw.com/PAGES/Whipple.html. Accessed November 14, 2008.

CHAPTER 2:
FIRST STEPS INTO BEHAVIOR PROGRAMS

Team Members' Roles in Behavior Programs

Ask any health care team members why they work in veterinary medicine, and the answer is always the same: "I love animals." It is this love of animals that motivates them to work in a veterinary practice, and it is also what keeps them there. Veterinarians and the entire health care team influence clients. Incorporating behavior programs into the practice is a logical role for non-veterinary professional team members.

Veterinarians should continue to focus on diagnosis, prognosis, prescription, and treatment. Non-veterinary health care team members are responsible for a host of activities to support the veterinarians and patients. They are the logical individuals to first learn and then practice basic behavior techniques. In addition, their time costs less than that of the veterinarian when answering client questions about basic behavior issues. This elevation of importance of non-veterinary professional team members in the clients' eyes provides the health care team with additional self-confidence and support. It confirms to them that they have chosen the right job and practice for their career. In the long run, they can see the results of their labor and realize greater satisfaction with their jobs.

The Three Levels of Behavior Counseling

Incorporating behavior counseling into a practice is not difficult and can be very rewarding, both emotionally and financially. Most practices perform some sort of behavior services anyway—they just do not realize it. This may occur through a puppy housetraining chat with an owner or in the course of other informal conversations with the client. Formalizing a behavior program allows the practice to emphasize its importance, develop the consistency needed to educate clients, and help the 72 percent of clients who ask their veterinarians about behavior problems.[1] Behavior programs also turn common pet behaviors into a revenue source.

Basic Behavior Program

Companion animal practices see puppies and/or kittens. Puppy and kitten packages incorporating vaccines and intestinal parasite examinations or treatments are very common. Beginning a basic behavior program starts with giving a behavior questionnaire to the client while in the reception area. Incorporate the PuppySmarts DVDs, AAHA brochures, and appropriate client handouts from this book into your existing puppy and kitten system, depending on the behavior concerns of the client. If you have materials that are already working for your practice, continue to use them. These client education materials should always be prescribed for the client by the veterinarian. Materials can be complimentary or offered for a fee as part of your puppy and kitten package, or may be prescribed as issues arise. Be careful, however, not to overextend your practice at this stage by giving out handouts on behaviors your team is not ready to address with the clients.

Having a successful program in your practice requires preparation, team members who are supportive, and a practice that fully learns and supports the process. Consistency is essential for success. If you use the DVD to teach the command *sit*, all the team members should teach *sit* in the same manner. The goal is to add value to clients' visits. All puppies and kittens up to 12 months of age should participate in the basic program because, again, we are trying to identify challenges before they become problems.

The basic behavior program requires a level of trust among the team members and a willingness to learn and provide a consistent message to clients. It is recommended that practices implement the basic program first before deciding to grow to the other options. A strong foundation is essential before proceeding to the next levels.

Intermediate Behavior Program

The next program level opens the opportunity for team members to become more involved with clients. New patient orientations are offered for groups of clients with puppies and kittens. These orientations are led and conducted by the non-veterinary professional team members but scripted by the veterinarian. The topics covered do not require veterinarian participation and can include basic behavior, health, Care-Credit, pet insurance, microchips, grooming, and so on.

It is best to have two or more team members working together to implement the program in a practice. They should work with the doctors to determine the topics to cover and how much time to spend on each topic. Such a program can significantly increase a practice's efficiency by moving time-consuming information dissemination from the initial patient visit into a group setting, making it a non-veterinary yet professional function. This also adds client benefit and value. This intermediate-level program allows time for the practice to introduce the advanced program when the entire team is ready to do so. Orientations can have a nominal registration fee.

Advanced Behavior Program

The advanced behavior program should be implemented once the basic and intermediate behavior programs have proven successful. One-on-one training sessions allow the maximum benefit to clients, patients, and the practice. The best advanced program implements Patient Behavior Advocates (PBAs), who are team members selected to be trained as PBAs. The role of the PBA is to provide additional information and assistance to clients in Patient Behavior Sessions (PBSs) that explicitly handle behavior challenges in young pets before the challenges become problems. PBAs see patients by prescription or client request. Because each puppy and kitten owner is unique, this advanced program schedules one-on-one visits that are separate from the medical wellness exams. These behavior-focused visits contribute to the mental and behavioral health of the pet and owner. This program will differentiate your practice in a major way and should be heavily promoted. A fee should be charged for these services.

The Step-by-Step Approach

This section explains in detail the steps to take in implementing the programs and gives in-depth suggestions on how to implement them. Keep in mind as you begin to execute these programs into your practice that change is difficult for many people and that it affects people differently. Some team members will adapt rapidly and embrace the changes being made; others will be slow to change or may be resistant and unwilling to change. For optimal success, it is important that all members of the practice are consistent in implementing the changes to which the practice commits.

Review the steps that follow for implementing behavior in the practice. Discuss and decide with your team what would work best. To implement these steps smoothly, it is very important to set clear goals. Follow up frequently with weekly meetings to review the direction the program is taking—what progress is being made, what needs to improve, and what may need to be done differently. Listen carefully to your team members to understand their challenges and issues. Let them know they are a valuable part of implementing this behavior program. You must understand where your team is presently before taking them where you want them to go.

At each step in the process, go over the progress made with the team members, celebrate success, and collect suggestions for improvements. Begin the next step when the team members are ready.

Step 1: Get Agreement

The first step in implementing the approach outlined in this book is to have a practice meeting that will create awareness of the problem that a behavior program can address. For example, you can cite that an estimated 9.6 million animals (56% of dogs and 71% of cats) that enter animal shelters are euthanized each year.[2]

Explain that most clients look to veterinary practices for expert advice on behavior problems just as they do in other areas in which they have concerns for their pets. Implementing a behavior program will give the practice the tools it needs to address clients' concerns. Ask how many team members would like to make a difference in the lives of their patients. Describe the program outlined in this book and its step-by-step approach. Get agreement from the team members to begin implementation.

Explain the role of the PBA, and define the protocols you want PBAs to follow. Ask for volunteers who want to become PBAs. Ask the potential PBAs to read this book and to meet once a week to discuss what they have read. This will prepare them for implementation of the PBSs.

Refer to the section on Sample Protocols at the end of this chapter to see sample protocols to use to implement each step of a behavior program. Each practice will want to set up its program slightly differently from other practices, but these samples will give you a start. If these protocols work for you, you can find them on the included CD, save them to your computer, and modify them. After you create specific protocols for the PBAs to follow, review and discuss the protocols with the PBAs before beginning Step 5 of the program.

Step 2: Measure Your Success

It is important during the implementation of this program to know when you are being successful. This will allow you to do "course corrections," if necessary, along the way.

One way you can do this is to gather historical information about the new puppy/kitten patients seen at the practice five months ago, for example, or even a year ago. Review how many of these new patients are still with your practice. One measurement you can use is how many of these patients completed their vaccine series. Some practices have found that their drop-off rate was considerably higher than expected. Some practices have found that even clients who prepaid for their puppy's vaccinations, spay, or neuter did not always return for those services. We tend to remember the patients that come back while forgetting the ones that "dropped out." You probably do not have access to the reasons why those who dropped out are not with your practice, but for a significant portion of them, the puppy/kitten has lost its home. Surrender statistics are a sad truth.

Review Chapter 1 on the business side of behavior. This will give you additional insight on exactly where your practice is and where you want to go. Set a goal for the practice. For example, in six months we want to see a 10 percent decline in the number of lost patients.

After implementing each step of this program, check your metrics against your starting point. Review what is working and what is not. Make adjustments as necessary to help your team members be successful before proceeding to the next step in the program.

Step 3: Basic Behavior Program

Step 1 involved deciding as a practice whether to implement behavior programs. It also identified who would play key roles in making such a program a success. In this initial step, you made decisions on protocols to be followed. Step 2 entailed developing a method of measuring the success of the program. Now, in Step 3, you implement the basic behavior program.

Preparation begins by gathering the tools to support the basic behavior program. These tools include the Puppy Behavior Questionnaire, version 1 (Appendix A); the Kitten Behavior Questionnaire, version 1 (Appendix B); PuppySmarts DVDs; AAHA brochures; and other materials you may be using now.

Version 1 of the puppy and kitten behavior questionnaires are designed for the basic behavior program and incorporate PuppySmarts DVDs and AAHA brochures in response to behavior challenges the client is having. It is important when making changes to these questionnaires that your questions are in "yes (true)" and "no (false)" format to keep the form easy for team members to quickly review and highlight. It is also important to have these training tools on hand to refer to if the client identifies a behavior concern. The specific topics covered by the PuppySmarts DVDs and the AAHA brochures are outlined below:

- PuppySmarts Training DVDs
  - Potty & Crate Training (Housetraining & Crate Training)
  - Biting
  - Jumping
  - Chewing (teaches the *leave-it* command)
  - Basic Obedience (teaches the *pay attention, sit, down, come,* and *stay* commands)
- AAHA Brochures (Canine)
  - Basic Training: Teaching Your Puppy to Mind His Manners
  - Busy Dogs Are Good Dogs
  - Crate Training: Creating a Canine Haven
  - Destructive Doggies: Solving Chewing and Digging Problems
  - Fearful Fido: Helping Your Dog Overcome the Fear of People
  - Home Alone: Solving Separation Anxiety Problems
  - Noisy Canines: Solving Barking and Growling Problems
  - Piranha Puppies: Keeping Mouthing and Biting under Control
  - Pushy Pups: Preventing and Understanding Aggression in Dogs
  - The Social Scene: Introducing Your Puppy to the World
  - Taking the Hassle Out of Housetraining Your Puppy
- AAHA Brochures (Feline)
  - Destructive Cats: Solving Scratching and Chewing Problems
  - Litter Box Blues: Solving Feline House Soiling Problems
  - Scaredy Cat: Helping Kittens and Cats with Fear
  - Taking the Hassle out of Housetraining Your Kitten
  - The Feline Terrorist: Taming the Kitten with an Attitude

Review version 1 of the puppy and kitten behavior questionnaires and the tools you selected with the team members so everyone understands the tools' purposes and processes. Everyone should watch the videos, read the brochures, and then discuss them. Understanding which tool will be appropriate for each client's and patient's needs will make it easier for the veterinarians to prescribe the best products for their patients.

A Description of Puppy's/Kitten's First Visit in a Basic Behavior Program

When the client registers at the front desk, the receptionist gives the client a copy of the Puppy or Kitten Behavior Questionnaire, version 1, and the new client paperwork. After completing the forms, the client returns them to the front desk. A team member picks up the patient's chart and escorts the client and patient to an exam room. The team member quickly reviews the behavior questionnaire and highlights any areas of concern. The team member performs normal medical procedures.

All team members should make the puppy's/kitten's first visit to the clinic as enjoyable as possible. New patients can easily become frightened if not handled gently and respectfully. This first visit has the potential to set the tone of the practice's relationship with your patient for life, so make it a positive one. Most young animals are curious, so allow them a few minutes to explore this new environment while gathering information from the client. Let young patients explore this strange new place with all its different smells and sounds. To make that first visit a pleasant, non-threatening experience, you may consider providing toys in the exam room for puppies and kittens, or using toys that incorporate treats to distract them while you are working with them.

Puppies and kittens will exhibit their concern by (1) turning to look at you; (2) turning to look at what you are doing; (3) yelping, crying, meowing loudly; and/or (4) licking and/or biting you. Stop what you are doing for a moment if the new patients exhibit any of the above behaviors. Pet them, soothe them, and offer them toys or food treats to distract them.

Do not pull kittens out of carriers. Instead, open the carrier and let the kitten walk out on its own while collecting information from the client. A toy with a ball or string top that sits on a stand placed approximately a foot outside the carrier door can catch a kitten's interest and may help entice the kitten out of the carrier.

The team member records any observations about the puppy's or kitten's behavior and interaction with the client and then leaves the room. Then the veterinarian comes in and performs a medical examination. The veterinarian reviews the information on the behavior questionnaire with the client and prescribes the tools carried by the practice for the specific behavior.

Since housetraining is the number one reason for surrender, always prescribe the PuppySmarts Potty & Crate Training DVD to every new client with a puppy as a "courtesy gift" during the first appointment. In addition, the AAHA brochure *The Social Scene: Introducing Your Puppy to the World* is vital to the overall emotional well-being of the puppy, so that, too, should be sent home with every new puppy patient. By doing these two things, you will help many clients settle their new puppies into their families. Whether you charge clients for either of these tools will be a decision made by the practice.

Once the appointment is complete, a team member will escort the client back to the front desk and relay the veterinarian's instructions to the cashier and thank the client for visiting the practice.

The cashier gives the "courtesy gift" to the client along with other information scripted by the veterinarian. Place the gift on the actual invoice with a charge that is discounted to zero as a "courtesy gift." For a puppy, this gift would be the Puppy Smarts Potty & Crate Training DVD and the AAHA brochure *The Social Scene: Introducing Your Puppy to the World.* For a kitten, it would be the AAHA brochure *Taking the Hassle out of Housetraining Your Kitten.*

What to Do If the Client Needs More than the Basic Behavior Program

Since you have not yet incorporated PBSs, if the client adds a behavior issue in the last section of the questionnaire, you can use the client handout from this book that addresses that specific behavior problem to help educate your client (see Appendices C through F and the included CD). You can also refer the client to a local behavior professional. You could refer to any of the following behavior professionals, based on the needs of your patient and client.[3] These resources represent professional help that is available outside the practice for serious cases that your practice isn't ready or able to take on, at this point or, perhaps, ever:

- *Board-certified veterinary behaviorist.* Board-certified veterinary behaviorists specialize in clinical animal behavior. They are qualified to diagnose and treat medical and behavioral problems and can prescribe medications to treat those problems. They are knowledgeable in the sociologic, psychological, and medical aspects of animal behavior. A board-certified veterinary behaviorist or Diplomate of the American College of Veterinary Behaviorists (DACVB) is a veterinarian who has completed the certification requirements of the ACVB, including a two- to five-year behavioral medicine residency, published a research study in a scientific journal, and passed a rigorous board examination.

- *Certified applied animal behaviorist.* A certified applied animal behaviorist has an upper-level degree such as a master's degree or a doctorate in an animal behavior-related field, such as psychology, biology, animal behavior, and behavioral ecology, and has met the educational, experiential, and ethical standards of the certification program of the Animal Behavior Society (ABS).

- *A veterinarian whose practice is limited to behavior.* A veterinarian whose practice is limited to behavior is a veterinarian who sees only behavior cases. He or she may have completed a residency program or may have taken additional continuing education courses on the subject. Only veterinarians can legally diagnose and treat medical and behavioral problems in animals, including prescribing medications to treat those problems. Many veterinarians who have an interest in animal behavior are members of the American Veterinary Society of Animal Behavior.

- *Behaviorist.* The term "behaviorist" is not attached to any specific qualification or level of schooling unless preceded by the words "veterinary" or "certified applied animal." The designation behaviorist can be used by anyone, including someone with no formal education or experience in companion animal behavior.

- *Dog trainer or pet behavior consultant.* No training or experience level is necessary for someone to use "dog trainer" or "pet behavior consultant" to describe themselves. A voluntary certification program for pet dog trainers exists through the Certification Council for Pet Dog Trainers, however, including a written examination, references from a veterinarian and fellow trainer, and proof of hours logged teaching.

Step 4: Intermediate Behavior Program

Once the practice has implemented the basic behavior program from Step 3 and the team is comfortable, you are ready to move to an intermediate behavior program. This step adds puppy/kitten orientations.

During your first appointment with a new puppy or kitten, you have a lot of information to cover in addition to performing the initial physical examination. You most likely cover the same things over and over with each client. You may already use a checklist to make sure you have covered everything. It has been suggested that practices demonstrate 17 different behaviors and/or training tips for puppies in the first visit alone! Meanwhile, clients are focused on their new puppy/kitten and not on what you are saying or showing them. The average person can retain only 5 percent of what you tell them without distractions. This rises to 10 percent if you tell them and give them written information to support what you told them.[4] You can only imagine how little of the information you give to your clients during their first visit is actually retained, understood, and acted upon.

Some of this information consists of absolute necessities, such as client sign-off on zoonotic diseases, vaccination requirements, physical examination results, and state and local regulations or requirements. Much of the information is important but not urgent, and can be handled in a new patient orientation. A puppy/kitten orientation is an hour-long gathering of a group of clients (without pets) who all receive this important information at the same time. Conduct these new patient orientations as often as you feel is necessary to meet your clients' needs. The orientations should be given by your volunteer PBAs or other team members selected for the purpose.

Once you are ready to begin offering puppy/kitten orientations, review how many new patients your practice sees in a week and in a month. Use this information as a way to measure the time the practice is spending with new patients versus the time you will be saving by implementing this program. As the practice becomes more comfortable with this program, you may need additional technicians. The veterinarians in your practice will have more time available for medical issues because they will not have to spend so much time with new patients. Best of all, clients will be able to pay closer attention and retain more information if their puppies/kittens are not with them during the puppy/kitten orientations.

When a client brings in a new puppy or kitten, the practice continues to use the Puppy or Kitten Behavior Questionnaire, version 1. The veterinarian will stress to the client during the appointment that she must attend a puppy/kitten orientation along with any other materials the veterinarian may prescribe. When the client goes to the front desk after the veterinary appointment, the cashier registers the client for the orientation. Any materials, such as the PuppySmarts DVDs or AAHA brochures that the veterinarian prescribed, are handled just as they are in the basic behavior program.

Listed below are just a few of the topics that can be covered in the puppy/kitten orientation. It is important that your practice's veterinarians add to or subtract from this list to deliver the information you would like your clients to receive. You probably already have certain information you give to your new puppy and kitten clients. Once your list is complete, you should formalize it by distributing it to your PBAs. Your PBAs should not veer too far from this script. Of course, remind your PBAs not to get in over their heads—it's always better to pass the buck than spread misinformation.

- *Preventative health products.* Products are available to prevent heartworm, flea and tick infestations, and diseases.
- *Microchipping.* Microchipping allows a lost animal to be identified and returned to his family.
- *Pet grooming needs.* Dental care, nail clipping, and good hygiene are necessary for a healthy pet.
- *Pet health insurance.* Applying for insurance when a pet is young and healthy provides a better opportunity for coverage when needed.
- *Medical expense credit card.* Credit preapproval for unexpected medical expenses (e.g., CareCredit) will eliminate the additional stress that your client may need to deal with in an emergency.
- *Spay/neuter.* Responsible ownership requires animal spay or neuter to minimize pet overpopulation.
- *Training needs.* The number one reason for the surrender of pets is behavior problems; being proactive in dealing with behavior problems saves lives.
- *Socialization needs.* Puppies and kittens are most receptive to socialization when they are young and curious, which will make them more comfortable in their environment.
- *Client expectations.* It is important that clients have realistic expectations regarding appropriate and inappropriate behavior. (Lassie and Benji are not representative of the average pet.)
- *Recommended puppy/kitten programs.* Programs that your practice offers or recommends will differentiate your practice and strengthen client loyalty.

As you begin to implement puppy/kitten orientations, have a weekly meeting with your PBAs to address their challenges or concerns. Find out what is working and what is not. Make adjustments along the way until the program becomes second nature to your PBAs and the practice. Once that happens, you are ready for Step 5 as you build on your practice's success.

As an illustration of the advantage of using PBAs for the orientations, let's consider a large practice that sees an average of 25 new puppy patients a day. They schedule a one-hour appointment for all new patients. The practice is open six days a week for 12 hours each day. This means that 72 hours of veterinarian time is currently spent with new patients doing a task that, for the most part, could be done by PBAs. PBAs can cover all of the nonmedical issues in the puppy/kitten orientation, making new patient appointments shorter, which means more billable hours for the doctors. In this particular practice, PBAs could do one-hour puppy/kitten orientations twice a day, six days a week. That would equate to using 12 team member hours a week instead of 12 veterinarian hours a day. How many new patients does your practice see in a week, and how much time do you spend with each new patient?

Step 5: Advanced Behavior Program

Step 5 is the advanced behavior program, which implements one-on-one PBSs. Steps 1 through 4 have led to the point where your practice and PBAs can have the maximum beneficial impact on your patient's behavior needs. Steps 1, 2, and 3 prepared your practice to begin a behavior program and got you started. Step 4 moved the communication of common information to a group setting, reducing time spent on a repetitive task. The PBAs are now in a position to take partial responsibility for client education. Before beginning this step, the practice should decide whether to begin with Puppy Behavior Sessions or Kitten Behavior Sessions. Get comfortable with doing training sessions for one species before adding the other.

Create a standard of care statement for your practice as it relates to behavior. You can find a sample in the following section, Standards and Protocols for the Advanced Behavior Program. You can adjust the sample standard of care to match the conditions in your practice and its existing standards.

Step 5 replaces version 1 of the puppy or kitten behavior questionnaires with version 2 (see Appendices A and B). Review the questionnaires and make adjustments as you did with version 1. Decide which behaviors you want your PBAs to address and which tools you will be using for each behavior.

Detailed sample protocols to implement the advanced behavior program are included in the next section as a suggestion for your practice. Review these protocols and make adjustments for your practice.

A suggested Puppy Behavior Session Checklist is included in Appendix A and a suggested Kitten Behavior Session Checklist is included in Appendix B (for kittens) as guidelines for PBAs to follow in gathering more detailed information about specific behavior issues. Review and make adjustments based upon the behaviors you want your PBAs to handle. This checklist probes the behaviors identified by the puppy/kitten behavior questionnaires.

The tasks above are best done by the practice's veterinarians before bringing the results to the entire team. Hold a meeting for the entire team to review the standard of care statement, updated puppy or kitten behavior questionnaires, the patient behavior session checklists, protocols to be followed, and the tools selected by the veterinarians. Get input from your team and from the PBAs. Set a date to begin implementation that allows the PBAs to become comfortable with the information in this book relating to the behaviors you want them to address. The volunteers selected to be PBAs in Step 1 must be reasonably familiar with most of the contents of this book before their first PBS. They can review the specific behavior concerns before each appointment, but a good general knowledge in advance will make a huge difference.

Put all of the materials, questionnaires, checklists, standards of care, and written protocols for the behavior program in a common location. Review them frequently in staff meetings until they become second nature to your team.

Standards and Protocols for the Advanced Behavior Program

Creating a Standard of Care for Your Practice

A standard of care is a statement of the desired level of care you want to give to your patients. This statement helps the practice establish a consistent, high level of care and defines the values that the practice is committed to deliver in the area of behavior. The following example of a standard of care statement can be added to your existing practice standards:

> *Every puppy/kitten visit (for patients 12 months of age or younger) will include a patient behavior questionnaire and a prescription for (1) training tools or (2) a Patient Behavior Session that will help clients address training and behavior challenges.*

Protocols Established by Veterinarians in the Practice

When reading the sample protocols on the following pages, you will have some decisions to make. You may have a qualified and competent PBA, in which case, you may allow this person to make certain decisions, which in turn will allow you to spend more time on other procedures. However, you may also feel that your PBAs should consult with a veterinarian before straying from your practice's pre-established behavior training protocols. Decide how the PBA will address and handle the following:

- How much discretion is afforded the PBA?
- Should the PBA discuss behavior problems with patient's veterinarian prior to each session or simply follow the notes in the patient's chart?
- How will referrals be handled? Is it recommended that veterinarians make outside referrals specifically to a board-certified veterinary behaviorist?
- How will referrals to outside trainers or others be handled?

- What behavioral signs warrant a physical examination before addressing behaviors?
- When should the puppy/kitten be re-examined by the veterinarian before a PBS is allowed?
- Who will communicate to the client that additional appointments will be necessary based on the PBA's evaluation at the end of each session?
- How should the PBA handle additional or different behavior challenges he or she is seeing while in the training sessions?
- How will the PBA address animals that are distressed when left alone (possible separation anxiety)? Do you want your PBAs to use the forms for this behavior in this book? Will you use medication to assist in addressing this behavior?
- How will aggressive behavior issues for puppies and kittens be addressed in your practice?
- Which veterinarian will come into the PBS to prescribe the appropriate tools and products at the end of the session?

Sample Protocol: First Visit to the Practice

This protocol is followed for all new puppy or kitten patients less than one year old on their first visit to the practice. Adjust this to match your selected standard of care statement. *Note*: All notations, observations, behavior prescriptions, and recommendations should be included in the patient's record and be initialed or signed by the person making the entry.

1. The receptionist greets the client and patient and welcomes them to the practice.

2. The receptionist gives appropriate (puppy/kitten) Patient Behavior Questionnaire, version 2, and new patient forms to the client. The client completes the forms while in the waiting room.

3. The client returns the completed forms to the receptionist.

4. The receptionist sets up a new patient chart.

5. A team member greets the client and patient and welcomes them to the practice.

6. The team member escorts the client and patient to the exam room while observing the relationship and interaction between them, and records these observations into the patient record.

7. A technician reviews the Patient Behavior Questionnaire form filled out by the client, highlighting "yes/true" areas of concern, and notes observations. This step will give the patient time to acclimate to the new environment in the exam room.

8. The technician performs established medical protocols (temperature, weight, etc.) and confirms client's concerns from Patient Behavior Questionnaire.

9. The technician reviews the medical and behavioral information with the veterinarian. Some patient challenges may relate to physical causes that may otherwise not surface. This questionnaire now becomes part of the patient's medical record.

10. A veterinarian performs a physical examination consistent with established medical protocol.

11. The veterinarian tells the client to attend a new puppy/kitten orientation.

12. The veterinarian reviews the Patient Behavior Questionnaire and makes a prescription for training tools, such as handouts from this book, the AAHA behavior brochures, or the PuppySmarts training DVDs or a PBS.

13. If the doctor prescribes a PBS, he or she notes what specific behavior challenge(s) the PBA should address in the session. (Note: A PBA can address only one or maybe two behaviors in any 45- to 60-minute session.)

14. If additional behavior challenges need to be addressed, additional PBSs should be prescribed to the client. The practice may want to offer a package of three to five PBSs.

15. The veterinarian tells the technician that the client should receive a new patient welcome gift (AAHA brochure and/or training DVD).

16. A team member escorts the client to the cashier's desk and communicates the veterinarian's prescription for training tools to be sent home and/or an appointment for a PBS. (Example of verbiage: "Lucy, Dr. Smith would like you to set up a Patient Behavior Session for Taffy, and give Mrs. Jones the Potty & Crate Training DVD as a gift. Will you please take care of that?")

17. The team member reminds the client and cashier to schedule puppy/kitten orientation, communicates to the client that it was a pleasure to meet them both, and excuses self from the desk.

18. The cashier schedules the new puppy/kitten orientation.

19. The cashier schedules the PBS, if prescribed by the veterinarian, and discusses options and fees for services with the client.

20. The cashier includes any additional fees on the invoice and gives the client the training tools the veterinarian has prescribed.

21. The cashier gives the "courtesy gift" when prescribed and posts a charge on the invoice. The cashier zeros out the charge for the "courtesy gift" as a goodwill gesture. This will show the value of the gift to the client and will begin to build the client/practice bond.

22. The cashier communicates to the client and patient that it has been a pleasure meeting them both and he or she is looking forward to seeing them both again at their next visit.

23. The cashier schedules a phone reminder for the PBA to follow up with the client for a status update in one week if a PBS was prescribed but not scheduled.

24. A follow-up call is scheduled for two weeks if any behavior problems were noted during the initial visit.

Sample Protocol: Telephone Follow-up after the First Visit[5]

Note: The phone call and any new information gathered should be included in the patient's record and be initialed or signed by the person making the entry.

1. One week after the first appointment, the PBA or appropriate team member places a telephone call to the client to check the patient's progress. (Example of verbiage: "Mrs. Jones, I know your next scheduled appointment for Taffy's vaccine is not until [insert appointment date], but the doctor and I wanted to touch base with you to see how Taffy is doing and ask whether you had any questions that we may be able to assist you with at this time. If not, great. We look forward to seeing you again on [insert appointment date].")

2. If the client is having behavior challenges, the team member will convey the information to the veterinarian.

3. The veterinarian may decide that an examination for possible physical causes, a PBS, or a referral to an outside source may be necessary.

4. This follow-up call is made by the same person who made the original call.

Sample Protocol: Preparation for the Patient Behavior Session

1. The PBA reviews the Puppy/Kitten Behavior Questionnaire, version 2, that the client filled out at the last appointment.

2. The PBA reviews the veterinarian's notes on what behavior(s) he or she wants addressed in this session.

3. The PBA reviews the behavior overview(s) the doctor wants addressed in the PBS. Puppy and kitten behavior overview(s) are in Chapters 4 and 5, respectively.

4. The PBA reviews the sample copy of the appropriate Puppy Behavior Session Checklist (Appendix A) or Kitten Behavior Session Checklist (Appendix B) that explains the purpose of each question and references appropriate materials and/or sections of this book. The PBA prints out a copy of the part of the checklist for use during the appointment. In addition, the PBA prints out a copy of any applicable client handouts to share with the client during the appointment.

5. During the PBS, the PBA should address only one or two behavior issues so as not to overwhelm the client, patient, or PBA.

6. The PBA should not ask all the questions on the Puppy or Kitten Behavior Session Checklist. Each behavior has a heading, so the PBA should use the relevant sections.

Sample Protocol: Guidelines for Conducting the Patient Behavior Sessions

Note: All notations, observations, behavior prescriptions, and recommendations should be included in the patient's record and be initialed or signed by the person making the entry.

1. The recommended time for a PBS is 45 to 60 minutes.

2. The PBA addresses only one or two behavior issues per appointment. Attempting to do too much will only confuse the client and patient.

3. Only veterinarians can diagnose and treat behavior disorders.

4. The veterinarian decides which behaviors the PBA is to address with the client.

5. The PBA follows all protocols set up by the veterinarian.

6. The PBA must always be mindful, patient, and respectful of both the client and patient.

7. The PBA does not use hurtful or forceful techniques when working with any behavior challenges the patient may present.

8. When doing PBSs, the PBA has appropriate products available for the veterinarian to prescribe, based on what the patient is experiencing. The PBA asks the veterinarian to step into the session just a moment to prescribe products.

9. The PBA reminds clients that toys patients can ingest, get caught in their claws, or tangle themselves in, such as plush toys, soft rubber toys, and rawhide, should be allowed only when the patient is supervised.

10. The PBA must explain how to use or fit any training tools prescribed.

11. If, in the opinion of the PBA, a concern requires additional expertise, or if the PBA feels uncomfortable working with a behavior, the PBA is to excuse himself/herself from the session and consult with a veterinarian to get directions on how to proceed.

12. The PBA will use the Puppy or Kitten Behavior Session Checklist as a guide for the PBS. If the patient's behavior leads the appointment in a different direction than planned, that is perfectly fine. The PBA should follow where the patient's displayed behavior leads.

13. The PBA must obtain medical clearance before proceeding with any behavioral advice that has not been recommended by a doctor prior to the appointment.

14. The PBA always asks the questions in the general section of the Puppy or Kitten Behavior Session Checklist. The answers will provide important information about the patient's background.

15. The PBA reviews the information in the client handout(s) with the client and, whenever possible, demonstrates how to train the correct behavior. The PBA asks the client to repeat what the PBA just did. A little coaching is often necessary.

16. The PBA has high-value training treats (e.g., chicken or cheese, dried bits of fish, something patients will *love*). Puppies and kittens will work very hard for great-smelling treats they really want, especially if they have not eaten prior to the appointment.

17. The PBA also has people treats available to reinforce the client's proper training techniques. Pet parents appreciate treats, too—Pretzels for diabetic clients and chocolate for others.

18. The PBA remembers to acknowledge when the client and patient do something right.

19. The PBA takes the time needed to make sure the client understands how to work with the behavior properly. This may take the client a few tries until she gets it right, so be patient.

20. Once the client seems on track, the PBA remembers to congratulate them. Praise will help them remember what they have learned.

21. The PBA then can give handout(s) to the client as a reminder and as a guide to use at home. The cashier can give the handout(s) to the client when she is checking out, if that is easier.

22. When ending the session, the PBA always thanks the client for caring and investing the time to help the pet overcome a behavior challenge.

23. The PBA must keep a clear record of any guidance, training tools, observations, concerns, or information given to the client during the PBS. The PBA writes down what the client says to him or her in the record, remembering also to sign all entries. This is important for liability reasons and for future reference.

24. Multiple appointments may be required to address all client concerns.

Sample Protocol: Patient Behavior Session Follow-up Call

1. One week after each PBS, the PBA places a telephone call to the client to check the puppy's/kitten's progress. (Example of verbiage: "Mrs. Jones, I wanted to touch base with you to see how Taffy and you are doing with [behavior] and to ask if you have any additional questions that I may be able to assist you with at this time. If not, great. We look forward to seeing you at your next visit.")

2. If the client replies that she does not believe her pet is making progress, then the PBA asks the client what is happening. Once the PBA has the information, the PBA tells the client he or she will discuss the information with the veterinarian and get back to the client.

3. The PBA discusses the client's challenges with the patient's veterinarian, who makes recommendations.

4. If the doctor recommends another PBS, an appointment for a physical examination, and/or testing, the same PBA that made the initial call communicates this to the client in a follow-up call.

5. The PBA records the information in the patient's record and signs or initials the entries.

6. If the doctor feels a referral to a trainer who uses reward-based training or a board-certified veterinary behaviorist is in order, then the doctor should make the follow-up call to the client. The PBA can make the call if directed by the protocol established by the veterinarians.

7. If the client talks to the PBA about a new behavior challenge that has presented itself, the PBA will follow the protocols established by the veterinarians. If, based on the practice's protocols, the PBA has been given leeway to set up additional appointments, then he or she does so. If not, the PBA discusses the client's situation with the patient's doctor for guidance. The PBA follows up with the client accordingly.

Sample Protocol: Additional Patient Visits

The protocol for additional visits follows the format of the first visit and continues until the patient reaches one year of age. The need for additional training materials and PBSs will be determined by the veterinarian based on information provided by the client on the Patient Behavior Questionnaire, version 2, or based on observation by a team member. The only departure is that a new puppy/kitten orientation is not necessary unless the client has never attended one. Also, there is no need for another courtesy gift.

Sample Protocol: Five Months to Maturity Follow-up for Puppies

1. The practice will send a postcard to puppy parents when the puppy is approximately five to six months of age.

2. During this stage in the puppy's development, the adult teeth are settling in and the puppy will go through a strong chewing phase.

3. The postcard should emphasize the need for good hard rubber toys for the puppy to chew on and the importance of managing the puppy's environment (i.e., crating the puppy when left alone).

4. During this time of strong chewing urges, even if the puppy has been well-behaved and understands the rules of his new home, it may be important to confine (crate) the puppy when the client cannot watch him.

5. The PBA should read the section Understanding Puppy Development Stages in Chapter 4 for more information on changes puppies go through between six and seven months of age.

6. A PBS may be appropriate if the puppy parent is having challenges during this time. Many puppies are surrendered between six and eight months of age.

7. At six months of age, the puppy is also going through sexual maturity. A letter, e-mail, or postcard should let the client know she may need to go back to basics with her puppy. As the young dog grows, so does his need to test his boundaries. Behaviors the client may have addressed a few months ago can reappear and will need to be addressed the same way.

8. During this stage in the puppy's development, another fear period can appear anywhere within the six- to twelve-month range. It is important that the puppy not be reprimanded harshly, as emotional damage caused here can stay with the puppy for the rest of his life. This is a time for the client to go back to basics in training as if her puppy was eight weeks old. The same patience and understanding will be needed, even though the puppy is much older and bigger.

9. Between 12 and 36 months of age, the puppy is going through social maturity. Again, behaviors the client may have addressed much earlier can reappear and will need to be addressed again.

Endnotes

1. American Veterinarian Medical Association, 2003 Survey

2. www.americanhumane.org/site/PageServer?pagename=nr_fact_sheets_animal_euthanasia, Data on file, American Humane Society 1997-2007, Accessed November, 2007.

3. http://www.flvetbehavior.com/Web/FAQ-Behavior_Professionals.html, courtesy of Lisa Radosta-Huntley, Accessed March 24, 2008.

4. David A Sousa, *How the Brain Learns*, 2nd edition (Thousand Oaks, CA: Corwin Press, 2000).

5. Tom Catanzaro, *Promoting the Human-Animal Bond in Veterinary Practice* (Ames, IA: Iowa State University Press, 2001).

CHAPTER 3:
PATIENT BEHAVIOR ADVOCATE PET TRAINING PRINCIPLES

This chapter and the ones following are addressed specifically to the team members who have been chosen for the role of Patient Behavior Advocate (PBA).

Your New Role as a PBA

Each animal is an individual, and no two animals will react in quite the same way. What works for one may not work as well for another. The science portion of training gives you the tools you need to solve problems. However, as a PBA, you must also involve your heart and instincts when working with animals.

When you focus on an outcome or a technique, you are no longer listening to what the animal is communicating to you. When I am working with an animal, I "leave my ego at the door" because I am there to serve the client and patient's needs at that moment. For example, it is difficult to address a biting concern with a puppy if he is cowering under the client's chair. At that moment, I have to honor where the puppy is and deal with the fear concern first. The entire session could accomplish nothing if I was determined to deal with the biting first. Learn to observe and listen to what the patient is telling you and trust your gut level feelings. In most cases, they will not steer you wrong.

The client would not be there if she did not care about her pet. So thank the client for caring enough to invest her time and money in the well-being of her newest family member. Use patience and understanding when working with clients. Over the course of a few months, you will quickly become comfortable with many of the behavior challenges new pet parents face and begin to see the commonality in these behaviors.

Once you become comfortable with the information you will be giving to clients, you may come up with a few of your own ideas, and that is great. However, always check the protocol established by the veterinarians before adding any of your own ideas or methods. The methods and information contained in this book have been reviewed or edited by board-certified veterinary behaviorists. (If any changes are approved and made to this information, please share your changes and the results by e-mailing them to lindaw@FirstStepswithPets.com, so they can be shared with others.)

It is important to realize you do not have to have all the answers. No one has all the answers. Behavior problems can be very challenging sometimes; that is why we have trainers, certified applied animal behaviorists, and board-certified veterinary behaviorists to turn to for help.

Your practice should have a list of professionals to refer clients to when behavior problems arise that are out of your comfort zone. Check the protocols established by the veterinarians in your practice to find out how referrals are to be handled.

Puppies/kittens should always be physically cleared by a veterinarian before offering any behavior suggestions or guidance. Refer to the protocols established by the veterinarians on timelines to make sure you are within the parameters since the last physical exam.

If you think there may be a chance of a new physical problem since the veterinarian last saw the patient, seek guidance from the veterinarian before moving forward with the patient, or refer to protocols set by the veterinarians. When animals are in pain, they are not able to tell us. They may show their concern by displaying fear or aggression. Some behavior problems that could have physical causes include:

- Excessive house soiling could mean the animal has a urinary tract infection.
- Biting could mean the animal is in pain.
- Chewing could mean the animal is having trouble with his teeth.
- Aggressiveness or bullying might mean the animal is in pain.
- Fearfulness or shyness might also mean the animal is in pain.

Keep in mind that most pet parents just need some easy-to-follow guidelines to help them be successful. Give clients just one or two ways to work with a specific behavior and send them home with information such as a client handout, brochure, or video that will refresh their memories on what you talked to them about and demonstrated during their Patient Behavior Sessions (PBSs).

If the pet has many behavior issues, do not try to address them all in one PBS. Start with the behavior the doctor has recommended unless the patient takes you in a completely different direction. Focus on one training technique at a time when working with a behavior. Consistency is important. However, do not continue to use a method that is not working. If, after a week or so, the patient is not responding even though the client is following your directions correctly, then it is time to try something else.

Schedule additional appointments to address other behaviors with which the client needs help if approved by the veterinarian. Give the patient and client time to implement what you teach them in the first training session. Let them both have time to be successful, and then you can build on their successes.

When the patient has many behavior challenges, addressing all of them may feel overwhelming to the client. This can happen because the patient was frustrated and did not understand what was expected of him. Once the client and patient begin to work as a team, the information the pet is receiving becomes clear. In many cases, that is what it takes for the patient and client to both be successful. The good news is that when the patient and client are working as a team, "other" behavior challenges tend to resolve on their own. Then you are the hero and the patient and client live happily ever after.

Although it seems hard to grasp, the truth is that very few dogs work to please us. It is likely that their behaviors have been reinforced. Dogs are innately selfish creatures and work because there is something in it for them. That is difficult for people to accept, but once they can see that pets work for reinforcement and not because they "love" them, clients can start to see their pets are not spiteful or angry or bad.

Kittens are more unique and mysterious than any breed of dog I have ever met. Kittens like company but they are more independent. Kittens have a much shorter attention span than puppies. Many cat owners "free feed" their kittens, making it more difficult to use treats as a reward. These traits make them much more challenging to train, although they are extremely smart animals. When working with cats, you will want to keep your movements limited and slow. Clients should not "free feed" while training their kittens.

Cats are mischievous, cantankerous, mysterious, and a joy to work with! If you ever have the need for a lesson in patience, work with a cat. Respect them; ignore them and quickly they will come over to you. Coaxing works sometimes, but not as often as ignoring them does. The most important thing I can tell you when working with kittens is to keep your training session short. If you stop paying attention before the kitten decides your time is up, she will keep coming back for more attention and training. If you exceed the kitten's attention span, however, you will lose control quickly.

As a certified Tellington-Touch (T-Touch) practitioner, I am specifically talking about T-Touch when I mention touches and T-shirts in this book. Information about these touches appears in the section Alternative Methods later in this chapter. Kittens respond very well to touch, perhaps even better than puppies do. They seem to be so sensually aware of touch that they really enjoy and can learn a lot from touches. As an example, I was working with a veterinarian's cat that always hid when company came to the house. In fact, this doctor had friends who had known her for years and yet never had met her cat. She invited me over to do some touches on Kitty. She had captured her cat before I got there, and I used a towel to confine her a bit, but I held her very gently. I actually started off doing touches on the towel to not upset her more. In minutes, I could feel her whole body relax as she stopped squirming to get away. I worked on her for about five to ten minutes and then let her go. The doctor and I had lunch, and I then worked on Kitty again for five to ten minutes before ending the session. Three days later, the veterinarian called me to tell me Kitty came out of hiding last night for the first time when company was at her house. Being a veterinarian, she knew how unusual this was, yet she was personally experiencing the difference with her own cat. So when you are working with kittens, using touches may help.

Touches work beautifully with biting, aggression, fear, or settling any animal into a new home. However, this is not just about the touches—it is about a shift in your mindset. Once you become more comfortable with touches, they will give you an opportunity for better interaction with animals; an ability to give more respect, more empathy, more understanding, and more openness to their concerns; and best of all, a willingness to listen to animals at their level.

Another alternative that works for both kittens and puppies are T-shirts (see the section Alternative Methods later in this chapter). T-shirts give animals a better connection to their bodies and help to build confidence. In addition, T-shirts help animals to calm or settle down.

These alternative methods are included in this book as additional techniques you can use. Clients can also be shown how to do touches as a different way to connect with their animals. Keep in mind that the biggest reason people have pets is for companionship and love. Touches and T-shirts can deepen the relationship between clients and patients faster than any method I have ever seen or used!

Puppies and kittens do present many normal behavior challenges. Clients have many of the same training challenges. In time, you will see these patterns and it will get easier. When you catch problems early or can offer preventative information to help clients be successful, everyone wins. This does not have to be difficult. The overviews and client handouts in this book are in easy-to-read language. Give yourself the time you need to become comfortable with the different behavior challenges and build on your own successes. Over time, you will become comfortable and you will know you are making a difference in the lives of your patients. You will be a true advocate for the animals. Believe in yourself.

Training Principles for Puppies and Kittens

This section addresses how animals learn.

Associative Conditioning

Associative conditioning, also known as classical conditioning, involves the establishment of a mental link between two unrelated events. When two events occur together frequently enough, we begin to assume the two occur as one, even if there is no cause-and-effect relationship between them. As a result, one stimulus can be substituted for another.

Ivan Petrovich Pavlov unintentionally happened upon this phenomenon when he was studying the digestive system in dogs. While measuring the dog's amount of salivation, he noticed that the dogs would start to salivate when food was intro-

duced; this was quite normal. However, Pavlov then noticed something unusual: After a period of time, the dog would start to salivate when it first saw the technician. Pavlov decided to investigate this occurrence further. First he measured the amount of saliva created by a dog when food was introduced. In this experiment, the food is called an unconditioned stimulus, and the salivation is called an unconditioned response. This is because salivation is a normal reaction for a dog when food is introduced. In the next step of Pavlov's experiment, he rang a bell while introducing food to the dog. The bell is a neutral stimulus; this means the sound of a bell would not normally make a dog salivate. After a period of time, however, the dog would salivate whenever Pavlov would ring the bell. The bell had become a conditioned stimulus, and the reaction of salivation had become a conditioned response because a dog would not normally salivate when a bell was rung unless the animal had been conditioned, or taught, to do so. To put it simply, the dog came to associate a ringing bell with receiving food. This was an extremely important and influential experiment in understanding the concept of learning for animals as well as humans.

Operant Conditioning

Operant conditioning is also known by many other labels, such as "cause/effect" or "action/consequences." Operant conditioning is a term coined by B. F. Skinner. Skinner sought not only to connect behavior with a response, but also to distinguish that a behavior is repeated because of a consequence of that response. To study his theories, Skinner created what is known as the "Skinner box." This is an extremely well-known experiment that is still used today to study animal cognition. In its most basic form, an animal is able to freely move within a box. A device that is easy for the animal to operate dispenses food. When the animal operates the device, it receives food. For example, a pigeon could be flitting about its box when, by accident, it taps the food dispenser with its beak and receives a pellet of food. Eventually, the pigeon learns that every time it taps the food dispenser (behavior), it will receive a pellet (reinforcement). This is the basis from which reward- or consequence-based training stems. (See the definition of reward-based training in the section on Training Terminology later in this chapter.)

By using operant conditioning, you can teach a puppy or kitten what the consequences are (good and bad) for each action, behavior, or event. There are five possible outcomes for every action, behavior, or event in a puppy's or kitten's world:

- Something good happens (positive reinforcement).
- Something good stops (negative punishment).
- Something bad happens (positive punishment).
- Something bad stops (negative reinforcement).
- Nothing changes.

Examples

In training, we need to make use of the animals' natural responses to these outcomes to mold their behavior. The following examples show how both classical and operant conditioning can be useful in training.

Example #1 (Classical Conditioning/Association)

Many puppies think food treats are a good thing. But they do *not* know our language, and the word "yes" means nothing to them. Try this: At variable intervals, say the word "yes" to a puppy/kitten and then quickly give the animal the treat. After a short time, the animal will begin to respond to the word "yes" as if it were the treat. This is the process of associating a neutral stimulus ("yes") with an unconditioned stimulus (food treat) that naturally results in the unconditioned response (puppy's/kitten's happiness). The treat naturally elicits the reaction of happiness. What you are doing is making the word "yes" (without the treat) elicit the response of happiness. This can also be achieved with a clicker instead of the word "yes." See the next section, Training Methods, for explanations of verbal, target, and clicker training.

Example #2 (Operant Conditioning/Consequences)

A puppy bites his mom too hard when nursing or playing. The mother dog will instantly correct the puppy. The puppy begins to learn how to control his bite through the results (consequences) of nursing or playing too hard. The mom added a strong bark, walked away, or snapped at her puppy to reduce a behavior (biting too hard).

Behaviors that are rewarded will increase in intensity, frequency, and duration. Rewards are anything the puppies/kittens enjoy or perceive of as value to them, such as toys, treats, pets, belly rubs, a game of fetch, and so on. However, sometimes certain behaviors are rewarded unknowingly and cause more confusion for a pet. For example, paying attention to a puppy/kitten by talking to him can be considered a reward, but scolding him can then be misunderstood and be considered a reward as well because it conveys attention. This type of attention can confuse a puppy/kitten.

Behaviors that are ignored will decrease in intensity, frequency, and duration. Not rewarding a behavior can be accomplished by stopping play, walking away, confining the animal, or simply not paying any attention to the animal. For a client to remove his or her attention from the puppy/kitten, a short time out in the crate may be necessary in some cases.

Training Methods

Verbal Training

This type of training consists of saying a word such as "yes" whenever a puppy/kitten does what is expected of them. Coupled with a reward, this reinforces the behavior that the client wants. Perfect timing is not necessary because the animal can also pick up on the emotional content of the mark ("yes" in a happy voice). For example, if you give the cue *sit*, when the puppy's rear hits the floor, you mark it with "yes" (happy voice) and give the puppy a reward. To understand the concepts of mark and reward, refer to the entries Mark, Reward, and Reward-based Training in the section on Training Terminology later in this chapter.

Target Training

This type of training is about using an object the puppy/kitten learns to touch with his nose. For larger breed puppies, the client can start off by using the palm of her hand. Instruct your client to place her hand right in front of the puppy's nose and say the word "target." With smaller breeds and kittens, the client can use a targeting stick or a long-handled wooden spoon. If the puppy or kitten does not reach out to touch the object or the client's hand, the client can try rubbing a little cheese or tuna on her hand or on the targeting stick or spoon. She can also put the treat between her fingers, sticking out just a bit for the puppy or kitten to try to reach. This will usually get the animal's attention, and he will touch the hand or spoon to check out the smell or get the treat. When he touches it, the client should say the word "target," to mark the touch and then reward the touch.

This method of training is great for getting the puppy/kitten from point A to point B without pulling, dragging, or placing the animal into position. Targeting is also helpful in training a puppy or kitten to do tricks and getting fearful animals to go through confidence courses. The target gives the puppy/kitten something to focus on, especially when fearful of new experiences.

Clicker Training

A clicker can be used instead of your voice to mark a requested behavior. Timing in this method is extremely important because there is no emotional content. It requires the trainer to click the second the animal performs the desired behavior (e.g., the second the animal's rear hits the floor when training the cue *sit*).

One way to help clients get the timing right is to give them a tennis ball and a clicker. Ask them to bounce the tennis ball, and the second the ball hits the floor, to click. Have them practice their timing until they can click the moment the ball hits the floor.

Some clients love clicker training because it is something new and different. Always consider that fact when talking to clients. Whatever it takes to make training fun for both animal and human that is safe is always a good thing.

Training Terminology

This section introduces the terminology this book uses in a logical, building-block order. For example, to properly define positive reinforcement, we must first define the terms "positive" and "reinforcement." Or to learn about jackpots, we must first know what a reward is.

Positive versus Negative

A *positive* occurs when something is added to the puppy's or kitten's environment. A *negative* occurs when something is removed from the puppy's or kitten's environment.

Reinforcement versus Punishment

A *reinforcement* is anything that causes a specific behavior to be exhibited more frequently. A *punishment* is anything that reduces the likelihood of a behavior occurring. Adding the concepts of positive and negative, we get the following definitions and examples:

- Positive reinforcement (desired behavior): Adding something good to increase a behavior. (Example: Puppy parent plays with her puppy. Something good, puppy parent's attention, is added.)
- Positive punishment (unwanted behavior): Adding something bad to decrease a behavior. (Example: Puppy receives a shock from his collar when he begins barking. Something bad, a shock, is added in response to his barking.)
- Negative reinforcement (desired behavior): Removing something bad to increase a behavior. (Example: Puppy receives a continuous shock while barking, but it stops when puppy becomes quiet. Something bad, a shock, is stopped when dog stops barking.)
- Negative punishment (unwanted behavior): Removing something good to decrease a behavior. (Example: Puppy parent walks away or ignores puppy. Something good, puppy parent's attention, is taken away.)

Positive reinforcement and negative punishment, when administered properly, are the two most humane principles in training. These two principles combined are the basis for reward-based training. They work this way: The puppy/kitten performs a desired behavior and gets a reward (treat/petting/loving). This is an example of positive reinforcement (adding a good thing to increase behavior). In contrast, the puppy/kitten performs an undesired behavior and gets ignored by the parent walking away (removing a good thing, their attention). This is an example of negative punishment (removing something to decrease a behavior).

Extinction

When an animal no longer exhibits a behavior, that behavior is considered to have become extinct. *Extinction* occurs when a behavior has been eliminated. This is the desired result of overcoming behavior challenges.

Extinction Bursts

When an animal has been performing a specific behavior for a length of time and then the client no longer wants the behavior, the puppy/kitten will not want to stop the behavior. In fact, the pet may offer more of the behavior in its unwillingness to change. This burst of behavior prior to the extinction of it is called an *extinction burst*. This can be challenging for clients because they think they are doing the right thing to address the behavior problem only to find the puppy/kitten is exhibiting the behavior more, not less. This will continue for a short time and possibly grow in intensity until the puppy/kitten finally understands and gives up the behavior.

For example, let's look at a puppy who has been jumping up on the client and has been rewarded for this behavior with the client's attention (positive or negative). Attention of any kind is still attention. Now the client wants the behavior to stop. At first, the puppy will not want to give up the jumping behavior and, in fact, will do it more frequently. It is a burst of activity. This can be challenging for a client because she is working on trying to stop the jumping behavior and the puppy is fighting the change. The client is not getting the results she wants nor expects. However, once the puppy realizes the rewards for jumping are no longer there, the puppy will, in time, stop exhibiting this behavior.

Another example might give more clarity to the term extinction burst. Suppose you come home every night, unlock your front door, and enter your home. But today you come home from work, unlock the door, and the door does not open. You push on the door to open it, but it still will not open. You take your key out and try again and again to open the door, but it still will not open. You push on the door harder and harder because you know this door usually works, but for some reason, the door will not open. Disgusted, you finally give up and walk around to your back door to get into your home. What you did at your front door just before finally giving up and walking around to the back door was an extinction burst.

Help your clients understand this behavior before ending your PBSs. When clients see extinction burst behavior, it is actually a good thing. This is normal and a good indicator that the behavior will be stopping soon. Once explained upfront, clients will be more prepared if their puppies do not respond as quickly as perhaps they thought they would.

Mark

The reinforcement of a desired behavior must be given when the behavior is presented; however, it can be difficult to give the reinforcement immediately when the behavior occurs. For instance, if you call a puppy by name, and he turns and looks at you, unless he is at your side, it is difficult to reinforce the behavior at that exact moment.

Trainers associate something called a *mark* as a substitute until they can deliver the reinforcement (a treat/scratch/pet). The mark can be a word such as "yes," a click from a clicker, or any other event or word the client chooses that can be given the moment the behavior occurs, even from a distance. Conditioning a mark is important. One must be able to reinforce a puppy's or kitten's behavior immediately when the behavior occurs. The mark is used as a substitute for the immediate reward. At first, this association needs constant reinforcement; however, only occasional reinforcement is needed to maintain it.

This is the role classical conditioning plays in training: establishing the mark.

- Original association: "yes" ⟶ treat ⟶ results in reinforcement

- After conditioning: "yes" ⟶ results in reinforcement

Cue

A *cue*, once called a *command*, is merely a stimulus you can provide to an animal for which you want a certain response (behavior). A cue must be something over which you have control. Most dog trainers use verbal cues or hand signals. Body language can also be used as a subtle cue to elicit a behavior. Common verbal cues are "sit," "down," "come," and "stay." An example of a hand signal is pointing your finger at the ground for "down."

Reward

A *reward* is something that the animal cannot see that is of value to the animal. It can be treats, scratches, belly rubs, or a special toy. Rewards have to match the individual preferences of each animal. The best trainers have the best rewards.

For example, in a puppy class, a student complained that her puppy would not perform a behavior in class that they had worked on successfully all week using kibble as the reward. In class, however, with all the distractions, kibble was not a good enough reward. Yet, I could get the behavior immediately from the puppy. The difference? I had better treats. The student had kibble and I had cheese. If you were the animal, which treat would you respond to faster?

Have different levels of rewards based upon the difficulty of the task for which you are asking. Special performance should receive a special reward, or even a jackpot of rewards.

Jackpot

Jackpots are a bunch of treats all at once; however, they should be given to the animal one at a time quickly. Use them whenever an animal has done better than expected, advanced to the next level of a training sequence, or when the difficulty of the behavior requested has increased. You can also surprise the animal with a jackpot from time to time just to keep him hoping and guessing.

Lure

A *lure* is a reward the animal can see and will follow. You can use this reward to lure an animal into a position or an action you desire. Once the animal performs the desired behavior, you mark and reward the behavior. Introduce a cue you want the animal to associate with the behavior. Once the animal has made the association between the cue and the behavior, stop using a lure and rely solely on rewards. The difference between rewards and lures is that a reward is something the animal cannot see.

Shaping

Shaping refers to the gradual step-by-step learning of a new behavior by reinforcing successive approximations of the desired behavior. For example, training a puppy to *stay* is a good example of shaping a behavior. At first, a puppy will *stay* only for a few seconds. You must mark, reward, and release before the puppy breaks the *stay*. Over time, the puppy is taught to hold the *stay* longer and longer to get its reward. Eventually, the puppy will hold the *stay* for several minutes.

Reward-based Training

Reward-based training combines two types of operant conditioning (consequences): positive reinforcement and negative punishment. When training a puppy, always reward the behaviors you desire (positive reinforcement) and ignore or walk away from behaviors you do not want (negative punishment).

It is important to acknowledge everything the animal does correctly. Reward-based training is considered a superior training method because it is highly effective, easy for an animal to understand, humane, and most important, it is respectful to the animal.

Training Sequence

You should follow a *training sequence* when beginning to train a behavior. You must first establish a mark, and then the sequence involves using a lure to get the desired behavior. Mark and reward the behavior. As an animal becomes consistent in performing the behavior, a cue is introduced and associated with the behavior. Once the association is established, you no longer use a lure and just rely on rewards.

Over the years, I have learned that puppies learn very quickly. As a result, I introduce the lure and cue at the same time. For example, I lure the puppy into a sit position, and the second the puppy's rear-end hits the floor, I say the cue "sit" and mark and reward the desired behavior. The finished behavior looks the same. I just combine the lure and cue so that the association between the cue and behavior forms quicker. This also allows me to stop using a lure much earlier in the process. However, this method does not work for training all behaviors.

Reward Schedule

A *reward schedule* refers to the frequency and type of rewards you give a puppy or kitten for offering desired behaviors. As a behavior becomes more established, you vary the frequency of the rewards you give. For example, when first beginning to train a behavior, it is important to mark and reward the behavior every time. As you add different locations and distractions or ask for more complicated behaviors, you use higher-valued rewards. Over time, you will reduce the frequency of the rewards to maintain the behavior. Review the Rewards Schedule client handout in Appendix C for more detailed information.

Socialization

Socialization is the process by which an animal learns how to recognize and interact with the environment around it. A puppy/kitten must learn how to respond to other living things, events, and objects. Socialization introduces a puppy/kitten to the experiences he will likely encounter in his lifetime as part of a family. This may include:

- Visits to the clinic without being examined
- Meeting people of all ages, from babies to very old folks
- Children running, chasing, screaming, and playing
- People in different uniforms, such as firefighters, police officers, postal carriers, or delivery drivers
- Objects such as canes, hats, glasses, costumes, wheelchairs, and walkers
- Rides in a car or an elevator, and stairs
- Visits to parks
- Loud noises, such as vacuum cleaners, radios, televisions, trucks, planes, and motorcycles

Refer to the next section in this chapter, Socialize for a Confident, Well-Adjusted Pet.

Desensitization

Desensitization is exposure to a controlled stimulus in order to accustom the puppy/kitten to something or to reduce a negative reaction. When you desensitize a puppy/kitten to something that makes him uncomfortable, you want to repeat exposure in a controlled manner. In time, many puppies and kittens will learn to ignore whatever made them uncomfortable. Begin with mild exposure to the stimulus and gradually increase the frequency and intensity by using small steps. Give the puppy/kitten time to adjust to the increased stimulus. It is important to work at the animal's pace and not force acceptance. When you push this process too quickly, you cause flooding (see definition below).

For more on desensitization, refer to the section Desensitizing and Counter-Conditioning to Overcome Negative Reactions later in this chapter.

Counter-Conditioning

Counter-conditioning occurs when you pair something "good" that the puppy/kitten likes with something "bad" that it doesn't like. This is an effort to change the association the puppy/kitten has to the bad stimulus. Counter-conditioning is often coupled with desensitization when trying to help a puppy/kitten overcome an adverse reaction to a stimulus.

For more on counter-conditioning, refer to the section Desensitizing and Counter-Conditioning to Overcome Negative Reactions later in this chapter.

Flooding

Flooding refers to exposing an animal to whatever the animal is afraid of all at once until the animal is no longer afraid. Unlike desensitization, it is not a gradual process; it is one extreme dump of information. Just as the name implies, it "floods" the animal with a large amount of stimuli of whatever the animal is concerned about. Flooding is not a good way to work with fear or anxiety in animals. It can be traumatizing and may damage an animal psychologically and/or physically.

Socialize for a Confident, Well-Adjusted Pet

Socialization prepares a pet for life with a new family. The more positive experiences the puppy/kitten has, the more well-rounded and capable of handling life situations the pet will become. For an experience to be positive, it must be done at the puppy's/kitten's pace rather than at the client's pace. It is important not to pull or force a puppy or kitten to explore, touch, or accomplish anything. Instead, let the animal approach new things when it is ready; the client should try to make all new experiences fun.

Encourage the client to introduce the puppy/kitten to other animal species (cats, rabbits, horses, goats, and so forth) as well as other members of its own species. This includes not just members of the pet's own family, or the next-door neighbor's, but all types of other animals: big, small, young, and old. However, before a client introduces her puppy/kitten to other animals, the client should make sure that the other animals are properly immunized and safe for the new puppy/kitten to be around. It is also important that these other animals do not have a problem with young animals, or they will defeat the purpose of this interaction.

When the puppy/kitten is being introduced to other people, the client should never hold the pet. Instead, the puppy/kitten should meet the person at its own pace. If the puppy/kitten does not want to greet the person, do not force the animal into the experience. The client should just thank the person for his or her time and move on.

The client should try introducing her puppy/kitten to other people again and again until the puppy/kitten is willing to go up to the person and receive a treat that the client gave to the person to give to the puppy/kitten. The purpose of this exercise is to teach the puppy/kitten to be comfortable around other people.

Pushing, pulling, or forcing the puppy/kitten in any way defeats the entire socialization experience. Positive life experiences are necessary to socialize the newest family member. It is important to work at the animal's pace and build on successes. Pushing too hard or too quickly can ruin a pet as a family member. The puppy's/kitten's confidence builds over time as he becomes successful with new life experiences.

Building Confidence

Imagine for just a moment that you are afraid of heights. What if I bring you to the roof of a 20-story building, and place a 16-inch-wide plank going from that building to the 20-story building next door, and tell you to walk the plank to the other building? How would you feel about doing that? Would you perhaps feel fearful or apprehensive?

What if I put a collar around your neck and simply pulled you across to the other building because I could? Would that help build your confidence level?

What if you watched me run back and forth across the plank, and I made it look like great fun? Would you think I was weird, but be a little curious as to how I could possibly have fun doing such a scary thing?

What if I put bundles of $100 bills two inches apart on that plank that stretched all the way over to the other building? Would you consider reaching for the closest stacks of money, then venture out just an inch or so at a time? Would you be stretching your arm out as far as you could? (I would.) Would you consider this to be a worthy reward for your efforts? Do you think you might then be willing to give it a try?

Treats to most puppies/kittens are like money is to us. This is why knowing what motivates a pet is so important. If you were a millionaire, walking the plank for a few hundred thousand may not be worth it to you. But if you were having financial difficulties, and you could get your hands on a few extra hundred thousand dollars in a few minutes by crossing the plank to the other building, would you give it a try? Most of us would. The motivation is there for us to want to succeed. That is the same motivation we want puppies/kittens to have when socializing and training them.

How would you feel if you took the risk, walked the plank, and actually made it across to the other building? Would you be impressed and proud of yourself? Do you think you might have a bit more confidence?

That confidence is what we want for our pets. We want them to feel proud of their achievements and to gain confidence through learning that they are able to overcome new obstacles and achieve success in new situations. This newly gained confidence will help them handle new experiences throughout their lives.

Desensitization and Counter-Conditioning to Overcome Negative Reactions

Trying to get a puppy/kitten used to a new and scary object or experience may be frustrating. New objects are intimidating to some animals. When desensitizing a puppy/kitten to a collar, a leash, loud noises, or anything that may make it fearful, keep in mind that each step should be rewarded. Small and gradual steps will propel the animal into success and build confidence at the same time.

Begin small, and as the puppy/kitten gets accustomed to a stimulus, increase the frequency and intensity. After each increase, wait until they get accustomed to the new level before continuing. Proceed at the animal's pace. If you move too quickly, you might flood the puppy/kitten, which can be harmful. Remind your clients that their pets should never be forced to do anything. Forcing completely undermines the building of confidence.

An Example of Desensitization

A puppy in one of my classes was a little chihuahua named Baby, who was one of those shy/fearful young puppies. It took three whole weeks to get her comfortable with someone putting a collar on her! We introduced Baby to other things while we were waiting for her to be comfortable with the collar. I would invite this student to class a few minutes early each week so we could work alone with Baby a little at a time.

Once we got the collar up to her neck, we started to hold the two ends of the collar together with our hands for just a second then release it while offering her treats as a distraction. I had the client tie a small hankie around Baby's neck during the week as homework. By the third week, we got to the point where we could put one end of the collar through the other, and finally were able to close the collar while Baby stood quietly, waiting patiently for the rewards that were now given much further apart.

I mention Baby because she was very afraid of pretty much everything. She would not venture more than one foot away from her mom. We did not have to worry about Baby running away or the client losing control of her in class. In time, Baby even conquered the confidence course. Target training really helped her to focus when she was first introduced to it. Once she actually conquered the confidence course, her tail was wagging, not tucked between her legs. After a while, she would go through the entire confidence course, turn to look at her mom and me with pride, and prance back to her chair.

Puppies such as Baby can take a long time with every new thing we introduce them to, and sometimes it is difficult to be patient. It has been my experience, though, that being patient and working at the puppy's pace pays off and makes every second you spent with them worthwhile. Puppies will reward you as a great trainer if you give them the chance to learn at their own pace.

By the end of the classes, Baby was playing with other puppies, getting treats from everyone, and walking with a purpose. Baby was still tentative with new situations after class, but she also had many moments when, given the chance to accept something new, she did.

Baby was a tiny puppy and I could have just put the collar on her; trust me, it would have been a lot easier for me. But remember, helping a puppy is not about what is easiest or best for you. It is about what is best for the puppy. If we would have forced Baby to wear her collar, she would have become even more fearful instead of more confident. We would have suppressed her timid behavior, and she would then have tried more aggressive types of behavior such as growling or snapping to show her fear. When we push too hard, too quickly, animals will react negatively. Baby likely would have ended up as an ankle biter, an excessive barker, or an aggressive dog. In fact, she could have ended up becoming all of them! Instead, she grew into a relatively stable little puppy who in time became pretty confident. In addition, since the client played such an important role in working with her, their relationship grew closer as well.

An Example of Counter-Conditioning

As a puppy, Taffy, my long-haired dachshund, was afraid of loud noises. I took her down to a shooting range in the middle of the afternoon when just a few people were shooting. When we got out of the car it was quiet, so she was fine.

After a few minutes, a gunshot went off and Taffy cowered. I told her it was okay and offered her a treat. For the moment, she was distracted by the food. After a few minutes, I could see that one of the men was aiming his rifle, so I started offering Taffy many tiny, wonderful treats. The gun went off and Taffy cowered again, but only for a second because she loved the wonderful treats I was giving her.

After that, three men were getting ready to shoot their guns, so again I offered Taffy more treats. This time she jumped back a little with the first gun and ignored the other two, still distracted with the treats she was getting. I told her what a great little girl she was and we went home.

I repeated this training exercise with Taffy over the next couple of days. By day three, Taffy did not even respond to the guns going off. She was more interested in the treats than the guns firing. We stayed at the range for about 30 minutes each day. We played fetch and she received many wonderful treats.

The following day we went back to the shooting range, and it seemed like most of the police department were there practicing. At first, I was going to just go back home because I did not want to flood her with this training. However, 10 or 15 shots were fired while I was turning the car around to leave, and Taffy did not react to the noise. Being a trainer and always wanting to build on her successes, I noted she displayed no concern for the guns going off. I decided to stop the car and see how she would react to more guns going off before exiting the parking lot. Taffy just ignored all the guns going off while inside the car so I parked the car. After a few minutes, I decided to take her for a little walk with lots of treats in my pocket to reward the brave behavior I was hoping for.

Taffy cowered for just a second when about 10 guns went off at the same time. I quickly offered her treats and told her what a brave little girl I thought she was. Taffy turned her attention to me for more treats and, of course, I obliged; the rest is history. Taffy and I played together with one of her toys while guns in the background were rapid firing. Taffy ignored the shots completely and stayed focused on our game. Over only a few days, she had become desensitized to the loud sounds of gunshots.

A few days later, we were on a walk, and a motorcycle drove past us. Taffy watched the cycle go by and seemed quite okay with it. Her experience at the shooting range also helped her with loud thunderstorms. When the loud crackles of lightning and the boom from the thunder happened, she just ignored them.

The moral of this story is to expose the animal to stimuli slowly, use distraction, reward often, and build on the puppy's successes. When you build confidence, the animal may surprise you by handling other stimuli they were once afraid of. And this is all due to the time and patience you give to them.

Alternative Methods

Touch Work

The information provided below is in part from my training as a Tellington Touch (T-Touch) practitioner and additionally through personal experiences working with animals. It is not necessary to become a T-Touch practitioner, but understanding how to use some touches on animals can be very helpful.

When doing touches, always breathe and be mindful of what you are doing. Use fast circular movement at first with a rambunctious or high-strung animal and then gradually begin to slow down your circles. You must meet an energy where it is to take it to where you want it to go.

Use slow touches on shy, timid, and fearful dogs. You may even want to use the back of your hand to do the touches. Animals seem to instinctively know the difference between the front and back of a human hand and are less threatened by the back of the hand. Perhaps they know at some level that we cannot grab them with the back of our hands.

Slow touches should be done in a clockwise circular motion—no more than one and a quarter circles in any place at a time. When you do the touches, do not simply rub over the animal's coat. Instead, take the skin with you while doing the circular movements. Touches can be done randomly on the animal; a pattern is not necessary. Random touches that are easy to do and comfortable are usually what will work best.

Normal touches take two seconds; count 1001, 1002. Fast touches on hyper animals are only about a half second. If an animal does not seem to like what you are doing, change the speed, the pressure, or the location of the touch. Be creative, be flexible, and experiment. Pressure on the touches should be no more than the amount of pressure you would use on your own eyelids when your eyes are closed, unless you are certified as a Tellington Touch practitioner and understand other touches and when and how to use them.

When doing touches, pause frequently (every three to five minutes) to allow the animal time to process what they just experienced. The intent of touches is to give warmth, awareness, balance, self-confidence, self-awareness, self-esteem, and self-control to the animal, and it does. So you can understand why you would want to stop frequently and give the animal a chance to absorb what he or she is feeling.

Touch will work on kittens and cats as well as it does on dogs—even on horses and just about on any animal you can imagine. When working with long-haired animals, one method that works nicely is to hold a small amount of hair and make a circle while holding onto the hair near the skin or scalp. Then slowly release the hair by sliding it through your fingers, and continue a long, slow stroke even after you are no longer holding onto the hair. Cats especially enjoy long, slow movements.

T-Shirts

Using T-shirts on puppies or kittens may sound a little strange to you. I've used them numerous times and have gotten great results. You may be wondering why you would put a T-shirt on a puppy or kitten. The reason to use them is to help the animal feel more grounded. The T-shirt gives an animal a better sense of self, if you will. When on, a T-shirt can change a high-strung animal into a calmer animal in minutes—literally! I used to go to resale shops and buy a few old t-shirts with spandex in them to use on puppies in my classes that were either fearful or rambunctious. In just minutes you could see a difference in the animals' behavior. That is why there is a product on the market called the Anxiety Wrap.

There are a few things to know about using a T-shirt on any animal:

- You will want the shirt to have spandex or another stretch material in it so it will cling to the animals' body.
- You want to make sure the shirt does not pull or restrict the shoulders.
- It may be necessary to cut the arms of the T-shirt to make it more comfortable on the animal.
- Do not use a shirt with spaghetti straps; you want to connect with the animal's body as much as possible.
- The shirt you use should have a neck opening and short sleeves.
- With small dogs, you may want to use baby shirts.

- The shirt must go from the shoulders all the way to the end of the animal's rib cage.
- Work up the length of time you leave the shirt on gradually, especially with long-haired animals.

Start off with the animal wearing the shirt for 15 minutes, and gradually increase the amount of time in 15-minute increments over a period of time. The shirt can stay on for one hour or eight hours, depending on the animal.

With longer-haired animals, leave the T-shirt on for 15 minutes or less and take it off. The reason you cannot leave the shirt on for a long amount of time on long-haired dogs or cats is because it flattens the coat, and if you leave it on too long, the animal's hair will hurt. The feeling would be similar for you if you were to style your hair in a different direction than normal and leave it that way all day.

If you would like to give the T-shirt method a try, go to a thrift store to purchase them for demonstration purposes in the practice. The client may want to purchase something cuter for home use.

Bonding Moments™

The Bonding Moments™ program is about desensitizing the animal to being touched all over its body. At the same time, it is about giving clients an opportunity to notice changes in their pets that may need veterinary care. Best of all, it promotes a richer, deeper human-animal bond while letting the client know how much you and your practice care.

Most people simply pet their puppies unconsciously, paying little or no attention to the animal after the first few weeks when the newness of the puppy wears off. In contrast, bonding moments are a special time for clients and patients to stop what they are doing and bond to one another.

When the animal is comfortable with being handled, physical examinations and grooming procedures (e.g., ears, nails) will go much smoother. Clients will become more aware of behavior or physical changes. In many cases, clients will bring their pets into the practice much sooner when a change is noticed.

Refer back to the subsection on Touch Work earlier in this section. Doing touches will deepen the bond between the client and the patient. It will help the animal calm down for future visits as deeper understanding between animal and human occurs. The level of trust and respect the client will have with her pet will astound you and make your job much easier for future visits.

You can introduce this concept to clients with a comment such as, "You have such a wonderful puppy. I bet you could create some wonderful bonding moments with this little guy." In most cases, the client will ask you what you mean by that. This is your opportunity to create a "learning moment" for the client and make future visits to the practice easier on you and your colleagues. Now you have the client's attention.

Many clients do not even know when their puppies are hurt or sick until it is too late. You can create bonding moments and at the same time help the client to keep a close watch on her puppy's health by doing little things. For example, ask your client to have her puppy stand and gently run her hand over his entire body. In time, the puppy will love this special attention. Even just picking up his paws and running her finger between the pads, or picking up his ears and petting each ear softly inside and out, will let the puppy know the puppy parent cares. Tell the client to take a moment and look into the puppy's eyes and tell him what a great puppy he is.

The best part of creating bonding moments is that the client can do it whenever she wants, so it will always fit into her hectic schedule. Too many people miss this window of opportunity to create a warm, loving relationship with their puppies.

Some clients say it also helps them to relax when they are petting their puppies/kittens on purpose with long, slow strokes or circular touches as opposed to simply petting them. The pet will appreciate the special time spent with him and help your client relax after a hectic day at work. If the animal has gotten hurt or is not acting like usual when the client gives him this special attention, the client will know right away and can bring him into the practice for an examination. Prevention and early detection are key to a healthy pet. Remind your clients that the relationship they build now will last a lifetime.

Now, you may not want to use these specific words, but the point of the conversation should remain the same. The patient, the client, and the practice would be a lot better off if clients spent a few moments a day to be with their pets on purpose. Puppies/kittens who are used to being touched all over will be more comfortable with a health exam and getting their nails clipped than those who are only scratched behind the ears or patted on the back by their owners.

This is not everything you would like clients to do to prepare their puppies/kittens for veterinarian visits, but it is a wonderful start. Most clients do want that special bonding time with their pets and enjoy it. It sounds better than simply telling clients they should get their puppies/kittens used to being examined by the veterinarian while at home between visits. Creating bonding moments sounds more caring and nurturing to clients, and it is. It is about giving their puppies/kittens their undivided attention for just a few minutes a day.

The only rule here is that if the puppy or kitten pulls away at any time, the client should stop and give their pet a few minutes to relax and try again later. By doing this, the animal realizes he or she is being listened to and will trust the client more the next time the client tries to touch them in the same area.

The pet may pull away more frequently when the client is picking up the paws or running her fingers between the pads. Over time, the client will be desensitizing her pet to having his paws touched or having his nails trimmed.

Common Training Errors

When a puppy or kitten does not perform a requested behavior, it is not because the animal has selective hearing or is being hardheaded. When a puppy/kitten does not properly respond to a cue, a few things might be going wrong:

- *Unrealistic expectations.* The client may not have a good understanding of training or may have overly ambitious goals.

- *Repeating a cue more than once.* A puppy or kitten can become habituated to a cue when it is used too frequently. Is the puppy or kitten being asked for a sit, or for a sit, sit, sit? To a puppy or kitten, this repetition can be confusing. Remind the client that she is teaching the animal our language. Should the animal sit the first time she says, "sit," or the third time? Is the cue "sit" or "sit-sit-sit"? Which form of sit is the puppy or kitten supposed to respond to?

- *Not giving the puppy/kitten time to figure out what has been requested.* Animals need 30, maybe even 60 seconds, to figure out what is being asked of them when learning a new behavior. Many clients do not allow their pets this opportunity before repeating a cue.

- *Moving too quickly from using lures to using rewards.* The client should not rush from lures to rewards before a behavior is understood by the puppy/kitten.

- *Forgetting to move from using lures to using rewards.* The puppy/kitten responds only when there is a visible lure.

- *Forgetting to mark or reward the behavior when and where it happens.*

- *Rewarding the wrong behavior.* For example, a client asks for a sit, the animal sits and then stands up, and the client still gives the animal a reward. This will only confuse the animal because the client is reinforcing the wrong behavior.

- *Training in an area with too many distractions.* It is better to begin training new behaviors in a quiet place with little or no distractions.

- *Training more than one animal at the same time.* Each animal should be trained separately in the beginning until the behavior has been established. Once established, the behavior can then be taught to both animals at the same time.

- *Forgetting to generalize the behavior.* Because puppies/kittens do not generalize as quickly as humans, the client should train the same behavior in many locations over a period of time and introduce distractions as the puppy/kitten becomes more proficient. When a puppy/kitten is taught the cue *sit* in the kitchen, he may not perform as well in the living room unless the cue has been taught in the living room as well. When a puppy/kitten has learned a cue in four to six different places, the animal begins to understand that *sit* means to "sit in any location."

Other Training Basics

Length of Training Sessions

Training sessions for clients and patients working together should be short—lasting five to ten minutes. By keeping the lessons short, your client will be able to work with her young puppy's or kitten's attention span and fit the training into a hectic schedule. As the puppy/kitten gets older and the animal's attention span increases, the client can increase the length of the training sessions.

This does not mean, however, that a one-hour puppy/kitten class is not a great idea. In fact, a puppy/kitten class is a wonderful learning experience for both clients and patients. Whenever possible, these classes should always be recommended if someone you know and trust is offering them in your area. Of course, the best opportunity is for the practice to offer these classes, but for many practices it is just not possible.

Make sure to instruct your client that training sessions should always end on a positive note. If your client is asking for a new behavior from the puppy/kitten, and the animal is having a difficult time with it, tell your client to go back to something the puppy/kitten understands so that he can be successful before the session ends. Training sessions should be both *fun* and educational for client and patient. In this way, training sessions are something a puppy/kitten will look forward to and enjoy. It is a special time for a pet when he gets attention from his parent.

When starting a training session, it is important that the client be relaxed and pay attention to her animal. The client should begin training in a quiet place where there are no distractions. A hungry animal will be more interested in treats than one that has just eaten. As the puppy/kitten becomes successful in learning a new behavior, and all family members can get them to perform the desired behaviors on cue, then your client can begin adding distractions.

Involve the Family

For a puppy/kitten to respond to all family members, everyone within the family should take turns in the training. This is a great time for your client to introduce different rewards. They should find out the puppy's/kitten's level of interest in the different rewards being offered.

For example, an adult that spends a good amount of time with a puppy/kitten might get by with just pieces of kibble as rewards. However, a three-year-old might need more interesting treats to help the puppy/kitten pay attention, so the child can see positive results from his or her training sessions.

Children should train the puppy/kitten the same way an adult does, with short lessons, varying rewards, and always ending the session on a positive note. Children should never be left unsupervised with a puppy or kitten.

Compliance always seems to be an issue with pet owners. It is very important that the whole family is working as a team to train the new pet. In my experiences with families, the kids have more time and are actually the best trainers.

It can be a real challenge when one member of the family demands too much, too quickly, and too aggressively, from the puppy/kitten. If it is a man, it's an added issue because men tend to be more physical and have stronger, more intimidating voices. However, if another family member is more willing to let the puppy get away with more of the inappropriate behaviors, that can cause problems as well. Both attitudes can have negative repercussions on the pet, and both will confuse the puppy/kitten.

When one person is demanding or short-tempered, the puppy/kitten is not given much room for error in its training. This can result in a shy or fearful animal. Unfortunately, it can also end up causing aggressive behavior. The puppy/kitten may start treating others the same way it is being treated.

Also, if the puppy/kitten is not allowed to do something today and then *is* allowed to do it tomorrow, it is getting mixed signals, possibly from the same person. This will confuse the pet and result in frustration. This may manifest itself as overexuberance in the pet; it may also appear as anxiety, aggression, or even fear.

If possible, the best answer is to get the family to agree on how the puppy/kitten will be trained before they even get their pet. Help each family member make allowances until all are in agreement on the plan of action. Needless to say, this is not often possible. Usually, we hear about the problems after the new pet comes into the home.

Then there are the children to consider. Many, if given clear guidelines on how to train, will do great. Usually, from three to seven years of age is about the time when children can take an active role in training a puppy/kitten. For children from about eight to eighteen years of age, make sure to address being kind. Roughhousing and cruelty are unacceptable.

Babies, toddlers, and young children should never be left alone with animals. Sometimes even older children should not be left alone with the family pet. In most cases when no one is watching, the puppy/kitten will be blamed for any and all mishaps.

Whenever there is a new puppy/kitten joining a family, try to make an appointment for the whole family to come into the session together. It is best to do your training sessions with everyone at the same time. It will make your life and the pet's life a lot easier. This way, you are nipping a potential problem in the bud before it has even the slightest chance to flourish into a fully bloomed disaster. If the family's hectic schedules will not allow for one training session, then set up a second session for the other half of the family, if necessary, and charge accordingly.

Get the entire family to agree on a training plan or method. Decide whether clicker training or verbal commands will be used. Decide whether hand signals will be used and what those signals will look like. What will they teach the puppy/kitten first, second, and so forth? How will they teach it? Get agreement from each family member who will be involved in the puppy/kitten's training.

Tell the family to sit down together and watch a good training video or read a book so everyone will be going in the same direction with the puppy/kitten. While in the session, ask them to actually set a time that would work for all family members to do this. Help the family schedule at least a month of dates and times when the family will sit together to read or watch a training lesson.

To many of you, this may seem a little challenging, but it does not have to be. For example, if I was called in to help a family address a behavior issue, I refused to accept the client unless the whole family was willing to participate. I did not want to waste their time or my own with trying to handle a problem without the whole family willing to commit time to their pet. My thoughts were that if they could not give me an hour of their time to help train their puppy or kitten, what chance would I have that the family would follow my instructions? Insisting that the entire family participate in the training session was possible for me because I was doing the sessions in the clients' homes. In a clinical setting, getting the whole family into the practice during business hours will be more difficult.

With PBSs, if the whole family cannot attend a single session, schedule multiple sessions. Work with as many family members as you can. Send materials home for the family members who couldn't attend the session.

You are the professional, and in time you will know that you can help the family. Let them see your confidence. If everyone is not on the same page with the puppy/kitten training, you will end up with a lot of finger pointing: she does this, he does that, and the other will not listen. Manage the client's expectations and work with as many family members as possible.

The puppy/kitten will not learn what is expected without consistency. If the entire family is not on the same page, they may only confuse their pet. If the pet does not learn, then he will stand a good chance of losing his home. No matter how hard the pet tries, he needs consistency to learn anything about our human ways, expectations, and—unfortunately sometimes—demands.

Our animals want our attention, they want to please us, and they definitely want the wonderful rewards they will get when they do so. Be honest with the family, and give them the facts and the statistics, if necessary.

Your job will be to set up boundaries and guidelines for the family to follow. Here are areas to consider when helping a family train its new puppy/kitten:

- Get the family on the same page on how training works. Send home the client handout titled Training Principles (in Appendix C).
- Explain the difference between lures and rewards (the client handout Turning Lures into Rewards, Appendix C).
- Ask how many people are in the household.
- Ask the age of each person.

Now, based on age, recommend which family members can perform which activities safely with the pet. Ask the following questions and get consensus on the answers:

- Who will train the behavior first (mom or dad?), and then which children?
- Who will keep the water bowl filled with fresh, clean water and the bowl clean?

- Who will take the puppy for a walk—how often and when?
- Who will feed the puppy/kitten—how often, how much, and when (mom or dad first and then the children)?
- Who will brush the puppy/kitten—how often and when?
- Who will watch the puppy when it is outside of the pen or crate?
- How long will that person watch the puppy before it is someone else's turn?
- Who will be the next person to watch the puppy?
- Who will take the puppy out to relieve himself after each meal?
- Who will take the puppy out when he wakes up from a nap?
- Who will take the puppy out after or during play (depending on the length of the play period)?
- Who will clean the puppy/kitten's teeth—how often and when?
- Who will clean the kitten's litter box—how often and when?

By now, you should begin to understand how much you will want to break down the chores for each family member. After you do this a few times, it will go much faster.

After a while, you should develop a chart to assist in this information. This will make it easier for you and your clients. If you create a chart, give it to the veterinarians in your practice to review before you implement it.

Always seek advice and approval from the veterinarians for anything new you want to add to the program.

Adding Distractions

Once a puppy/kitten consistently performs a behavior, and other family members have trained the behavior, you can start to practice in new locations and add distractions or other elements. These new situations all make it more difficult for the puppy/kitten, but doing this helps the puppy/kitten to generalize the behavior.

For example, suppose a puppy/kitten is asked for a *stay* while a ball is introduced. At first, the animal should be asked to *stay* for only a few seconds; the client should then mark, reward, and release the animal from the cue. Instead of a regular treat, this time the puppy/kitten should get a special treat, such as a tiny piece of chicken or fish. The client should use a special reward for giving a desired behavior with distractions. This makes staying still worthwhile to the puppy/kitten because he just got a great reward! Next time, your client may want to tell her to *stay* and start bouncing a ball as a distraction. The client should give two, three, or four tiny pieces of meat or fish (jackpot!) if the puppy/kitten holds the *stay*. With a *stay* cue, always remind clients that after they give the reward they must release the pet from the *stay* cue.

CHAPTER 4:
PUPPY BEHAVIOR SESSIONS

Understanding the Puppy's History

Appendix A contains a Puppy Behavior Session Checklist to guide you with your Puppy Behavior Sessions. The general section of this checklist contains important questions about the puppy's background and his new family structure. The puppy's history will give you a better understanding of the puppy.

The first set of questions relate to where the puppy came from. This will let you know whether the puppy came from a reputable breeder, a pet store, a shelter, or a backyard breeder, or if the family found the puppy or the puppy found them. Each scenario may present unexpected challenges.

Pet Stores

Pet stores offer their own set of challenges for both puppies and new unsuspecting puppy parents. More than 90 percent of puppies in pet stores come from puppy mills! These puppies have had little human contact and have had little opportunity to have their emotional and developmental needs met. Puppy mill puppies generally live in a small space with their littermates with little or no stimulation or human contact until they are old enough to be sold.

For example, let's look at a puppy from a puppy mill who reaches the age when it can be sold. Up to this point, the puppy probably had very little human contact. In less than 48 hours, he goes from a stimulus-poor to stimulus-rich environment. The puppy was shipped to the puppy broker and experienced the following:

- A four- to six-step inspection by the broker
- Acceptance or rejection by the broker
- Examination by the broker's veterinarian
- Grooming to be made ready for shipment
- Shipment to a pet store anywhere across the country
- Examination by the pet store manager
- Isolation until the pet store's veterinarian comes to examine the puppy
- Examination by the pet store's veterinarian
- Being put in a window, crate, crib, plastic pool, or other unfamiliar container to be viewed and handled by people interested in purchasing him

On top of this, puppies are often handled inappropriately in pet stores. The longer it takes for a puppy to be sold, the more behavior problems the puppy is likely to have. In addition, these puppies are often sick, which has been shown to cause an increase in behavior problems when compared to puppies from professional breeders.

This initial stimulus-deficient environment followed by this stimulus overload can result in such behaviors as biting, fear, excessive barking, separation anxiety, and crate soiling, to mention just a few.

Despite these experiences, some puppies from puppy mills actually have an excellent disposition and adapt readily, so do not automatically assume anything. Watch the puppy and pay close attention to his body language.

Professional Breeders

Breeders who are aware and conscientious about their puppies' socialization needs usually have happy, normal, playful, and mischievous puppies. Any behavior challenges the client is experiencing with a puppy from a professional breeder are probably normal behaviors a puppy goes through, and the client just needs a little help addressing them. Utilize the section in the Puppy Behavior Session Checklist for the particular problem the client is having, and you should be well on your way.

In the search for the perfect "show dog" that will be a consistent winner in the ring, some breeders will take shortcuts that can emotionally harm an animal. Pushing too hard or too fast can cause a variety of behavior issues. If a dog develops behaviors that make him unsuitable for the showring, the breeders will try to sell the animal. Depending on the behaviors, the dog may have "special needs," which can make him more challenging as a family pet.

For example, suppose a puppy was five months old when the client got him from a professional breeder. The puppy comes from great stock, has beautiful markings, and exhibits excellent gait. The breeder believed this puppy could be the "perfect" specimen of the breed, but the breeder pushed the puppy too fast for the showring, and now the puppy is fearful. The breeder cannot successfully show a dog that is fearful, timid, shy, or aggressive in the ring, so the breeder sells the puppy. Meanwhile, your client thinks she just got a great puppy from excellent stock, and might even expect very high intelligence from the animal. It is best to approach behavior challenges with these clients gently.

Shelters

Consider the many reasons a puppy is at a shelter in the first place. The puppy may have been dropped off because the mother died during the birth and the owner did not want the responsibility of raising puppies. The litter may have been dropped

off because the owner did not want the litter. The puppy might have been found outside. The puppy may have had a home and has already lost it. In many cases, the puppy was not given the chance to be successful. For all you know, this puppy could have had two or three homes before your client's.

Or perhaps a previous owner may have had unrealistic expectations of the young dog, and when the puppy did not do *X* in *Y* amount of time, the puppy was relinquished. Many shelter puppies have a little extra "baggage" to deal with in the beginning. Many were surrendered because of unresolved behavior issues. The thing to keep in mind with these patients is their age. You may have to first undo a behavior before you can replace it with a new, more appropriate one. With a little time, patience, understanding, and guidance, these puppies can become some of the best dogs!

Many shelter puppies are very grateful for their new homes and try very hard to please their new families. They need and deserve patience and clear guidelines to follow to be successful in their new homes, so that this time, they get to keep their families.

Backyard Breeders

Backyard breeders range from excellent to poor. The backyard breeder may have raised the puppies in a stimulus-poor environment or a stimulus-rich environment. For puppies from backyard breeders, you could add a few questions to your questionnaire: Were there children in the home where the puppies were born? Did the children have access to the puppies? Was it supervised access? To the best of your knowledge, do you believe the puppy may have been abused or mishandled by unsupervised children or uncaring adults? How much were the puppies handled and socialized?

Some backyard breeders who are excellent handle and socialize their puppies by the book. The temperament of the sire and dam will be the best indicator of the puppy's behavior. When puppies from good backyard breeders are in a Puppy Behavior Session, the client may need just a little help to get the puppy on track. When appropriate, always ask clients if they met the sire and dam. If they did, ask about the temperament of the parents. This will give you a good indication of what the client and you can expect from the puppy.

Puppy's Age as a Factor

Another question in the general questions section of the Puppy Behavior Session Checklist refers to the puppy's age when the client acquired him. This can be very important information, along with where the puppy came from, in determining any "special" needs.

There are several questions to consider:

- Did the puppy have a chance to learn bite inhibition from littermates (did the puppy stay with the litter until he was eight weeks old)?
- Was the puppy confined until 12 or 14 weeks of age? If the answer is "yes," crate soiling could be one of the challenges the client is facing. By that age, the puppy has developed a preference on where to eliminate.
- Could this be a puppy the client purchased from a breeder after the puppy failed in the showring?

Read the next section, Understanding Puppy Development Stages, for some of the behavior issues that may arise if developmental needs were not met. The age of the puppy when acquired is very important.

Understanding Puppy Development Stages[1]

Neonatal Period: Birth to Two Weeks

The first period of the puppy's life is called the neonatal period. During this time, the puppy's main function in life is obtaining nutrition and staying warm. Puppies' eyes are fused closed and their ears are not open until 10 days old. Puppies can react only to what they can touch or taste. Since they cannot regulate their own body temperature, contact with mother or littermates is essential for their survival. This forms the basis for their behavior at this point. At this stage, their learning ability is severely limited and is connected to events that physically touch them.

Behavior is limited to making noise (care soliciting), investigatory movements (where's mom?), eating, and elimination. Puppies react to pain, cold, or hunger with the same behaviors—they make noise and move randomly. They do not appear to learn from mistakes at this stage. Their mother (or surrogate) has to stimulate them in order for elimination to occur.

Newborn puppies should be confined to a nest or whelping box both for temperature control and for continued contact with their mother and nest mates. Otherwise, their random movements could carry them far from their nest.

Lactating mothers produce pheromones starting at three to five days after birth. These pheromones are chemical substances secreted by the skin in the mammary area. The pheromones provide the puppies with a feeling of reassurance and comfort and help to ensure that puppies remain close by.

It is a good idea to pick the puppies up for just a few minutes a day during this time. A gentle massage on the head, legs, paws, and body can help the puppies physically and emotionally as they mature.

Throughout their development stages, touches play a wonderful role in helping puppies (as well as kittens) get a great start in life. It is my experience that the animals that receive touches will be calmer, be able to handle stress better, be healthier, and learn faster as they mature than their counterparts who have not received these touches.

Transitional Period: Three to Four Weeks

The transitional period refers to a puppy's huge physical development changes in its sense and physical abilities. Eyes are now open, teeth begin to erupt, and senses of hearing and smell develop rapidly. Puppies are beginning to stand and walk like miniature dinosaurs. They wag their tails and bark. Sight is well developed by four to five weeks of age.

At this stage, puppies begin to eliminate and defecate on their own. They start biting and pawing each other clumsily. Lapping and chewing begin, and they decrease their suckling reflex on objects that don't produce milk. However, their puppy teeth are effective only for soft foods.

During this period, the behaviors of the neonatal period begin to be replaced by more adult behaviors. Investigatory behaviors are beginning to focus on sight and sound. By three weeks of age, puppies can begin to orient on objects in the same room by sight instead of by touch.

Puppies removed from their mothers prior to three weeks of age are often emotionally and mentally stunted. In some cases, with touches, time, patience, and appropriate socialization, these puppies can be successfully integrated into some families. They may always have special needs, however.

Some puppies that are separated early can never be completely rehabilitated into being an acceptable pet for the average owner. It is a fine line. We want to offer the owner hope and get them interested in training because it will help, but we also do not want to offer false hope that they can make a puppy that has been separated too early into Lassie. That causes frustration and guilt on the owner's part.

Socialization Period: Three to Twelve Weeks

This stage is all about the puppy's social development. The puppy learns how to interact with his environment and members of his social group.

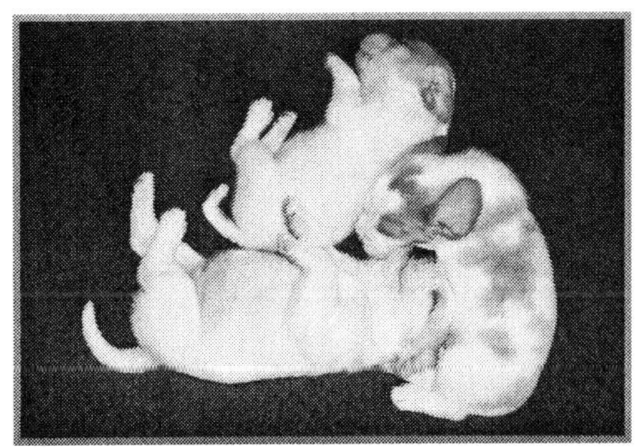

Weaning begins around the fourth week, and the client can begin supplementing the puppy's diet with pre-softened foods. By the fifth week, the mother dog may not lie down to nurse and is beginning to growl at the puppies when they chase her to try to nurse.

Early in this stage, the puppy will try to leave the sleeping area to urinate and defecate. At first, he will not go too far away, but he will try to eliminate away from where he sleeps and eats if possible. By eight and a half weeks of age, the puppy is already showing distinct preferences regarding where he will eliminate. The puppy wanders and sniffs, not because he is following a trail, but to find signs of urination or defecation that trigger his elimination response. This behavior is useful for housetraining. If the client uses a crate, the puppy is not likely to soil his sleeping area. If the client frequently shows the puppy the area she has chosen for him to eliminate and monitors him while he is awake and active, housetraining will go much smoother. If housetraining indoors, the client must be sure to properly deodorize all accident sites but leave scent traces in the acceptable area she has chosen, to trigger the elimination response. Puppies can eliminate every hour or two when awake and active until they are twelve weeks old.

Puppies also begin to vocalize in response to restraints (such as crates), peaking in frequency and volume when they are about six to seven weeks old and tapering off about the twelfth week. This is more a process of the puppy maturing than the puppy becoming "accustomed" to his crate.

During this period, puppies begin responding to sounds and sudden movements. Their learning abilities rapidly teach them which situations are dangerous and which have no meaning. Puppies may show fear of sudden movements of a human even if no prior bad experience exists.

Puppies normally become somewhat afraid of humans around the fifth week if they have not been handled, but with constant handling over the next two weeks, this fear can be almost completely eliminated. Research done by Freedman, King, and Elliot[2] has shown that a puppy's fear of humans increased the longer human contact was delayed. Puppies would investigate humans readily early in their development (three weeks or so), much slower at seven weeks, and not at all when older (fourteen weeks). It becomes apparent that puppies require early interaction with humans in order to become comfortable with handling.

"Play" fighting appears early during the socialization period (fourth week). In the fifth week, puppies begin to act as a group ("pack") rather than individually. A combination of this pack behavior coupled with play fighting can sometimes lead to group attacks. At about the seventh week, puppies can "gang up" on one of their littermates, often a small and weak puppy but sometimes a large or aggressive one. Scott and Fuller[3] noticed that in certain breeds, this play fighting can get out of hand, and human intervention may be needed to protect the victim.

Sexual behaviors such as mounting may appear as early as the fourth week. Other behaviors, such as beginning to sleep away from the nest, occur more often.

The age of seven and a half to eight weeks is the optimum time for puppies to enter a new home. They are active and ready to explore their new environment. At this age, they are most adaptable to new experiences. This is an excellent time to begin teaching them the skills they need for living with their new families. These skills include socialization, learning how to play properly with humans, and some basic manners.

Puppies removed from their mothers prior to seven weeks of age often are overly noisy, anxious, or fearful, are sometimes aggressive or bossy, have little bite inhibition, and exhibit poor canine social skills. Socialization, desensitization, and patience are absolutely necessary for these puppies to be successfully integrated into their new families. Consistency and clear guidelines for the puppy and client will be essential.

Research done at Purdue University[4] suggests that there is another "fear" period between eight and ten weeks of age. Some of the behavior modification techniques, such as "pinning" a puppy (holding the puppy down so he cannot move), during this time can enhance the puppy's fear greatly, stunting him emotionally and socially for life. Dominance techniques cause nearly irreparable harm and are no longer recommended.

Puppies must be frequently socialized with other dogs and humans to allow for species identification and development of appropriate human-dog interactions. Other species (such as cats or birds) must be included early (before the twelfth week) to inhibit predatory responses or fear of the unknown.

Many puppies today come from stimulus-poor environments, which can lead to poor social development. If these puppies are then placed into a stimulus-rich environment (such as a family with young children), the experience can be over-whelming. Emotional and behavioral problems, such as fear, anxiety, and even biting, can develop.

Puppies need to be exposed to safe, stimulus-rich environments to ensure their continuing physical and emotional development. But do keep in mind the environmental history of the puppy. Different opportunities to investigate are wonderful for the young dog. It is the client's responsibility to keep the puppy safe (e.g., from aggressive animals or people) and offer new stimuli at the puppy's pace. A puppy should not be rushed or forced into anything. We must all build on the puppy's successes.

Puppies born in puppy mills and sold at local pet stores in most cases will not receive the human contact they need during the important initial stages of their development. As a result, many of these puppies will be shy, fearful, and prone to biting. Many will need both you and the client to have more patience with them. With time, these puppies can be integrated into their new homes nicely. Sharing with a client the puppy's lack of social skills and how important that is to the young dog can help the client become more patient and understanding with her puppy.

Juvenile Period: Three to Six Months

Puppies stop growing as fast when they reach two-thirds of their adult size at approximately 16 weeks of age. Most of their physical development in this period is in terms of strength and skill. Dogs with more opportunities for physical development, such as exercise and play, will develop quicker.

Puppies kept in kennels until 15 weeks of age will resist learning to eliminate in new areas. They also have few inhibitions on eliminating near eating and sleeping areas and may readily soil their crates.

At 16 weeks old, the speed with which a puppy learns slows down. They have begun to form the patterns that control how they view the world. At this point, inappropriate behaviors have to be "unlearned" to replace them with new ones. Clients may experience extinction bursts (see the section Training Terminology in Chapter 3), when trying to stop a behavior that has been established.

Male dogs begin lifting their legs to urinate as early as the fifth month or as late as the eighth month. Some males may never lift their legs.

At about the fourth month, puppies may also go through another fear stage when they become easily afraid of unknown situations and/or people.

Adolescent Period: Six to Eighteen Months

This is the "teenage" period for the young dog. This is a time when the puppy will test the boundaries with just about everything it has been taught to date. The puppy will challenge other pets in the home for ranking and test family members as well to see how committed the family is to the rules the client thought were established. Knowing about this time in a puppy's life up front will make this early period go a lot smoother for both the client and patient. This is the time when clients who have not been forewarned will give up on their dogs and relinquish them.

With clear guidelines and consistency, the puppy will get through this period rather quickly. However, if rules are changed or the family lacks consistency during this time, all matters relating to the puppy may change.

Somewhere between five and six months of age, the puppy's adult teeth emerge. A puppy's need to chew will increase. During this time, puppies that once could be

trusted unsupervised in the home may now need to be put into a crate when unsupervised because the urge to chew is so great. A hard rubber toy for the puppy to chew on while in his crate is important and will satisfy his chewing needs.

It is generally accepted that a dog has achieved full adult status somewhere between 18 months and two years of age. During this period, the dog's behavior is most influenced by the members of its social group, both human and animal. If not spayed or neutered, sexual behaviors develop.

According to the research done at Purdue University,[5] another fear stage occurs during this period, which lasts for about three weeks. It is difficult to predict exactly when this will occur. As with all fear periods, if not handled properly, the emotional damage to the puppy could last throughout the dog's lifetime. Clients should ignore the moments of fear and reward brave behavior during this period.

Puppy Training Basics

Puppy Socialization

Review the sections in Chapter 3 titled Socialize for a Confident, Well-Adjusted Pet as well as Training Terminology (for the definition of socialization) before reading this section. The following information applies to puppies only.

From about seven and a half to twelve weeks of age is a crucial time for a puppy to be socialized. This is indeed the window of opportunity. Socialization can help all dogs of any age, but this early period is the easiest time to socialize them.

Some clients with smaller breed dogs carry their puppies everywhere. You probably have seen these puppies come into the practice without even a collar, let alone a leash, on. These clients may believe they are protecting their wonderful little puppies. What they are doing is telling their little puppies they are too small to handle anything on their own. If continued, the puppies may bark at other dogs, animals, and people constantly.

There are some special concerns for puppies when it comes to socialization. Puppies are sensitive to walking on different surfaces such as cement, grass, plastic, gravel, sand, and wood. When a puppy is first brought home is a great time to start introducing him to different textures, toys, and sounds. Taking a puppy outside when he is very young offers a chance to explore grass (weather permitting) and the opportunity to begin elimination outside. Clients should make sure the area the puppy uses for elimination is free of feces from other animals and has been treated for fleas and ticks.

Introducing the puppy to an old rug (as they may soil it), cement, and water on the floor for them to walk through all offer the puppy new experiences. Introducing safe rubber toys the puppy can work on to get a treat out will stimulate the puppy mentally.

Collars

Some puppies react poorly when you attempt to put a collar on them. It is normal for puppies to scratch their necks when they begin wearing a collar. It is something new they are not used to. If a puppy is reacting badly to a collar, you will need to gradually desensitize him to the feel of the collar, or the client may want to use a harness instead. If the client wants to use a collar, the following is one way of handling the situation in a Patient Behavior Session (PBS).

Let the puppy smell and lick the collar first. Then place the collar on the puppy's back to let him get used to the feel of it. While you are offering the puppy treats, ask the client to move the collar up the puppy's back toward his neck a little at a time. Once up to the neck, have the client remove the collar from the puppy's body and tell him what a brave little boy he is and give him lots of pets and treats. Show him the collar again and let him smell it. When he pushes the collar away to get to the hand with the treats, have the client put the collar on the dog while you offer the treats. That is usually all it takes; however, there are always exceptions. Read the client handout on Collars in Appendix D for more options.

Leashes

Leashes are another thing many puppies tend to dislike at first. You can appreciate their dislike when you realize it is our way of restraining them. Most creatures (including humans) do not like being restrained. Puppies must get used to wearing a leash, for in most cases it will be a necessity for life.

We want to teach puppies to love their leashes. Leashes represent wonderful things such as walks in the park and rides in the car. Here is a helpful rule: If your client does not want the puppy to pull her when on his leash, then she should not pull the puppy. Instead, have the client coax the puppy with wonderful treats or a favorite toy.

The client can also teach her puppy to *target* before introducing the leash. (See the section on Training Methods in Chapter 3 for a description of target training.) By doing this, the puppy will be able to perform a behavior with which he is familiar and comfortable. Training a puppy to target can be done in minutes. The puppy will not be proficient, but will know what targeting means. The only thing to remember is to move the target farther and farther away from the puppy slowly. Do not start at three inches away from the puppy's nose and then go three feet away. Instead, extend the distance the puppy will walk to touch the target gradually (e.g., three inches, six inches, one foot, and so on), and always remember to *mark* and *reward* when the puppy touches the target. (Refer to the section on Training Terminology in Chapter 3 for an explanation of *mark* and *reward*.)

Loud Noises

Loud noises, such as motorcycles, gunshots, thunder, loud music, garbage trucks, etc., can concern some puppies. If the noise does not bother the client, though, seven out of ten times it will not bother the puppy. If the noise *does* bother the client, then in many cases it will also bother the puppy. In general, an animal picks up emotional cues from its owner. The puppy will pick up on the client's reaction and act accordingly. Therefore, the client should always show confidence in what she is doing. The puppy will sense that emotional state and act accordingly.

As an example, I taught puppy classes at a community building that was part of a shooting range in Sunrise, Florida. It was amazing how some students would come in and actually be afraid of the gunshots. Almost instantly their puppies would start acting timid and afraid of the noise. Other students would come in and completely ignore the gunshots, even when the police department was shooting off 20 or 30 guns at a time. Most of those students thought it would be a good experience for their puppies, and it was: Most of these students' puppies were not concerned about the shots being fired either!

Exposing puppies to loud noises will help desensitize them and may prevent noise phobias in the future. The destruction that a thunder-phobic dog can do is legend. Some get sick, some tear up the house, and some actually jump through sliding glass doors or break down closet doors to get away from the sound.

Have your clients expose their puppies to a sound CD if they live in a quiet place. Just because it is quiet today does not mean five years from now they will not be living somewhere with violent thunderstorms, backfiring trucks, or Fourth of July fireworks. When your clients expose their puppies to these noises, have them simply play with their puppies as if everything is fine. If a puppy is apprehensive, have your client offer a few treats to get his mind off the noise. In a short time, he will learn to ignore the noise. If the puppy is still showing concern after a few minutes, the client should stop whatever noise stimulus she is offering. Do not flood the puppy with this new sound experience. Some puppies need time, patience, and many opportunities to become desensitized and conditioned to hearing loud sounds.

Exercise and Play

Exercise and play are important for all puppies. Exercise develops the puppy physically, and play helps develop the puppy socially and mentally.

Exercising a puppy is more than a walk on a leash or a jog around the block. Exercise consists of activities that build strength, agility, and coordination.

Likewise, play is more than letting the puppy run around the backyard alone. Play is the interaction between the puppy and other family members (or other animals). This is how a puppy builds social skills and learns how to interact properly with others. A puppy needs to learn how to play safely with people and what is and is not accepted behavior. Jumping, biting, and chasing are not acceptable ways for a puppy to play with people.

Many activities incorporate both exercise and play. They can be opportunities for training at the same time. For information on appropriate ways to exercise and play with puppies, read the client handout Exercise and Play in Appendix D.

Cues/Commands

Pay Attention

When training a puppy, the first cue the puppy needs to learn is his name. Other training is secondary. The puppy needs to learn his name so when you train him you can call him by his name to make sure you have his attention before you start to train a cue or other behavior. When teaching a puppy to *pay attention*, you want the puppy to look at you every time he hears his name. Once you have his attention, training can begin.

To begin training this behavior, take a good-smelling food treat (lure) and show it to the puppy. Once he smells the treat, say his name and slowly bring the treat up to your face between your eyes. The puppy will follow the treat up to your eyes. When he looks at the treat between your eyes, mark and reward him with the treat.

If the puppy does not follow the treat with his eyes, either you did not have his attention to begin with or the treat was not interesting enough for him to follow. Try again, and this time, make sure he sees the treat and you have his attention. Do this in an area with little or no distractions in the beginning. Repeat this exercise several times until the puppy is consistent at following the hand up to the eyes. Every time he does, mark and reward him.

The next step is to turn the lure into a reward. Use your hand the same way you did when you were holding the treat in it. Use the puppy's name and show your hand to the puppy just like you did when you had the treat in it, and bring your hand up between your eyes. The second he looks up at you, mark and reward with a treat that you have hidden in your other hand. Repeat this several times until he is consistent at following your hand up to your eyes.

Now try placing your hands at your side and calling him by name. Give him a few minutes to figure out what you want him to do. The second he looks up at you, jackpot him with many tiny treats. This was a big step for him to accomplish.

If the puppy did not follow your hand, stop working with the puppy and have a little chat with the client for a few minutes. Share with the client that in many cases puppies just need a little time to think about what we are asking them to do. When we leave the puppy alone for a few minutes to think about what just happened, many puppies seem to understand the behavior we are asking for a little easier.

Once you are successful with using the reward, give the client an opportunity to repeat what you just did. This way, when the client practices at home, you will know she is training the behavior properly.

Once the puppy is consistent, distractions can be added. If you ever have the opportunity to teach puppy classes, this is a great homework assignment for the first class. Having the puppy's attention prior to teaching him any new behaviors will make training new behaviors much easier.

The client will need to practice this with her puppy many times over a few days for the puppy to really get it. Once she has the puppy's attention, training any behavior will go a bit easier.

Send the client home with the client handout on Pay Attention in Appendix D, or you can send her home with the PuppySmarts Obedience DVD, where this technique is demonstrated nicely. That way the entire family can watch what the training should look like and help the puppy learn to look at family members whenever he hears his name.

Note: Remind the client not to overdo this. After all, we do not want the puppy to learn to ignore someone when they call his name. This training can be taught in sessions lasting just a minute or two, a few times a day.

Come

In the training world, the cue *come* is referred to as a recall. Most clients have never heard of a recall, so just use the word "come" when working with clients. The cue for this behavior is to use the puppy's name and say the cue "come"; for example, "Spot, come." When you are training the cue *come*, clients often make the following errors:

- The client calls the puppy by name, but forgets to say the cue "come." To train this behavior, the puppy's name and the cue must be used together. The name being used first is to make sure the client has the puppy's attention. The cue lets the puppy know what the client wants. In this case, the client wants the puppy to *come*.

- The client stands too far away from the puppy when she begins to train the behavior. The client should not be more than three feet away from the puppy when she begins training this cue.

- The client uses a low, flat, demanding tone of voice. This is neither exciting nor enticing to the puppy. Happy voices should always be used when asking a puppy to *come*.

- The client forgets to mark the first forward movement the puppy makes toward her. This marking is very important because we are teaching the puppy our language. The puppy has no idea what the word "come" means. He just knows the client has a wonderful-smelling treat, and he is trying to figure out what he needs to do to get the treat. When the client marks the first step the puppy makes toward her with a word like "yes" or a click from the clicker, the puppy will realize that forward movement is what the client wants. Repeating the mark a few times while the puppy is offering forward movement is a good idea, too. This is reassuring to the puppy, and many puppies will come even faster with the added encouragement.

- The client habituates the puppy to the cue *come* by overusing it. If the puppy is constantly being interrupted when playing or eating or sleeping, the strength of the cue *come* becomes diluted and the puppy learns to ignore it.

- The client forgets to change from lures to rewards. Over time, the puppy will only *come* when he can see or smell wonderful treats.

- The client uses the cue *come* when the client needs to do something the puppy may not enjoy or when the puppy is being scolded. To have a good recall, the puppy must be conditioned that only good things happen when he hears this cue and comes when called.

Clients should use different names for their puppies when calling them for different reasons. As an example, let's use one of my dogs named Gabrielle. When I want her to *come* quickly I use the cue "Gabby, come." When I need to clean her ears or trim her nails, I use the cue "Gabrielle, come here please." When she responds to "Gabrielle, come here please," she receives a very special treat, but she also knows something else is going to be happening to her. This is her slow recall. She will *come* because she has been conditioned to, but she will not be running over to me like she would if I gave her the cue "Gabby, come." "Gabby, come" always means good things are going to happen. "Gabrielle, come here please" means I want to do something to her.

When training a puppy for the cue *come*, it is important for the client to encourage him. The puppy should believe wonderful things are going to happen when he gets to the client. The client must be the most interesting thing in the world for the puppy. The client should talk to the puppy, use a high-pitched, excited voice, or clap hands in encouragement until the puppy reaches her. Once the puppy reaches her, the client must quickly mark and reward the puppy for coming on cue.

The easiest way to demonstrate the cue *come* in a PBS is by playing monkey (puppy)-in-the-middle. Ask the client to stand up with you and place yourselves three feet apart. Have plenty of treats ready for you and the client to give to the puppy. Give the client a few wonderful treats and have her show one to the puppy as a lure. You hold the puppy's leash first and then have the client ask for the *come*. Release the leash the second the puppy begins to move forward. Make sure the client marks the forward movement and then rewards the puppy with a treat when he reaches her. Once the treat has been given to the puppy, the client is to stand up straight and ignore the puppy. Now it is your turn. In a happy voice, say the puppy's name and give the cue "come." Remember to mark the first forward movement the puppy makes. Once the puppy comes to you, mark and reward the puppy. Stand up straight and ignore the puppy. Each of you, if possible, can take one step farther apart and repeat the exercise. Make sure the client marks the first forward movement from the puppy.

In a few minutes, the puppy will catch onto the game and run to the other person when called. The puppy may even begin running to the other person before he is called. Once you reach that point, whomever the puppy is farthest away from should be the person that calls the puppy. This teaches a very important cue, and it is a fun exercise for puppies and puppy parents alike.

Here is a trick I have used successfully for many years when older puppies just will not *come* on cue. Have the client go to the grocery store and buy a rotisserie chicken. Take this wonderful-smelling chicken off the bone and put it into a few plastic bags. In most cases, the puppy will be right there, watching and smelling the wonderful chicken and hoping for some to drop on the floor. Once the chicken has been deboned, have the client put it into the refrigerator for a minute while she puts a few of the puppy's normal training treats into her pocket.

Then the client should take one of the small bags of chicken out of the refrigerator when the puppy is not watching her. She can take the puppy outside to play. After five or ten minutes, have the client give the cue "Spot, come" in a happy voice (assuming the puppy's name is Spot). If the puppy looks at her and continues to play, have the client wait a few more seconds and repeat the cue in a happy voice while shaking the plastic bag with the chicken in it so the puppy can see it. The second the puppy begins to take forward movement toward the client, the client must mark the forward movement. Once the puppy comes back to the client, the puppy is given a reward of one of his normal treats and is released to go play again.

After another five or ten minutes, have the client call the puppy again, this time without showing the puppy any treats. If the puppy comes quickly, the puppy is rewarded with a piece of the wonderful chicken from the bag. If the cue must be repeated and the bag shown to the puppy, then again the regular treat is given as the reward. After a few tries, the puppy will begin to understand that when he sees nothing and comes quickly, he gets the chicken. If he has to be asked twice or shown the bag with the chicken in it, he gets a regular treat, not the chicken.

I have used this little trick with many older dogs who would not *come* on cue, including hunting dogs. It is amazing how quickly the dogs figure this out and will immediately *come* when called.

I was at a seminar with other trainers about seven or eight years ago, and all of us with dogs were staying at the same motel. After class the first day, we were standing in the parking lot deciding where to go for dinner. I suggested we let our dogs off-leash for a while as they, too, had been in class all day. There was a huge field behind the motel and I knew Gabby, then four years old, would love to run a bit. Everyone opted out, so Gabby had to run and play by herself. While we were all chatting, Gabby caught sight of a large heron, and the chase was on. All of a sudden, we realized there was a truck coming down the road between the fields that I had not seen. All the trainers in unison said, "Oh, my God," anticipating trouble. I called to Gabby by saying, "Gabby, come," and she turned around so quickly to come, she flipped over. She got up, and literally flew back to me. The other trainers were amazed that she had a recall that strong. I knew my little girl had a great recall or I would have not let her off-leash in a strange place. That recall was established when she was around three or four months old by using the methods I outlined above. To this day, at age 12, she has what I call a "flying recall."

Down

Some puppies are very reluctant to *down* on cue. This is a vulnerable position for a dog. First review the client handout on Down in Appendix A to see how this behavior is normally trained. If the information there does not work with some of your patients, here are a few additional ideas that may help.

Demonstrate for your clients the following simple exercise. Put the puppy on flooring he can easily slide on. Sit on the floor with your legs in front of you. Bend

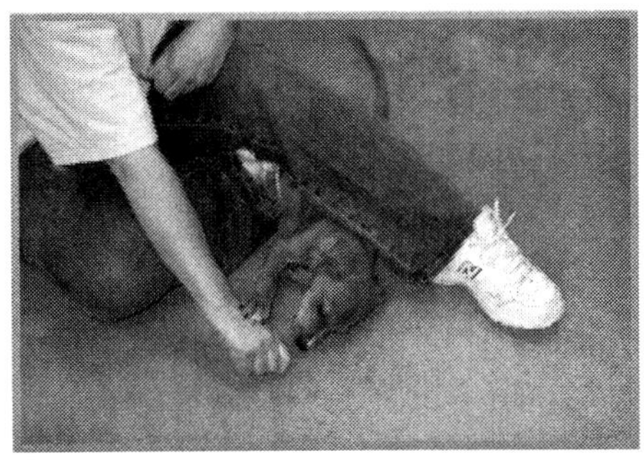

one leg up so your foot is completely on the floor. Then, show the puppy a treat by slipping your hand under your bent leg. Slowly bring your hand with the treat in it back through the bent leg, allowing the puppy to follow the treat with his nose. As you get closer to the bend in your leg, begin to lower the treat so the puppy can get under your knee to reach the treat. Then slowly bring your treat all the way to the floor and slide the treat toward your other leg. For the puppy to reach the treat, he will need to lie down. When he does lie down, quickly mark and reward him with a jackpot of treats for offering the *down*.

If getting on the floor is too difficult, sit next to a chair that has a horizontal rung on it. Use the horizontal rung as described above for your leg, and lure the puppy into a *down*. The most important part of this technique is to have the puppy on a slippery surface. The slippery surface will work to your advantage.

If the puppy still will not give the desired *down* behavior, ask for a behavior the puppy is willing to do and end the training session on a positive note. Wait a little while and try again. Many puppies need time to think about what just happened. Given five or ten minutes to think about it, many times the puppy will complete the task the second or third time you try.

Drop-it/Leave-it

A puppy can sometimes pick up things that are dangerous or that they simply should not have. The client needs a way to tell the puppy to release whatever is in his mouth.

All too often, clients grab items out of their puppy's mouth instead of training the cue *drop-it* or *leave-it*. In time, the puppy learns that when he is chewing on some-

thing and the puppy parent approaches, the chew item will be taken away. Then the problem of chewing on the wrong thing becomes a game for the puppy and the chase begins. Unfortunately, over time, if not addressed properly, the puppy will grow stronger and begin to defend what he has in his mouth. Then growling, snapping, or biting can occur.

To avoid these problems, the client can do a few different things: (1) puppy-proof their house, (2) train the cue *drop-it*, (3) train the cue *leave-it*, or (4) learn how to take dangerous items out of the puppy's mouth safely.

I have successfully used the cue *leave-it* to mean both *drop-it* and *leave-it*. At first, the idea may seem like quite a stretch, but it really is not. When you train a puppy to leave something alone, the puppy understands the item should not be put in his mouth. If the item is already in the puppy's mouth after the cue *leave-it* is trained, the puppy can quickly reason through the situation and release the item. Sometimes the old adage "less is more" even works in training an animal. Training either or both cues will make any item easier for the client to retrieve from the puppy's mouth without confrontations.

If you are uncomfortable using the one cue for both desired behaviors, training the cue *drop-it* is a game of exchange for items of higher value to the puppy. For example, suppose the puppy is chewing on a toy. The client walks up to the puppy with a chew toy that has some tasty treats inside and offers it to the puppy. The second the puppy drops the original toy for the great toy with treats inside, have the client give the cue "Drop-it." The client should mark the behavior and give the puppy the toy with the treats inside as the reward.

Another way to train *drop-it* is to have the client walk up to the puppy and offer him a tasty treat and give the cue "Drop-it." The second the puppy drops the toy from his mouth to receive the treat, the client marks the behavior with "Yes" or a click from the clicker and rewards the puppy with the treat.

In time, the puppy will learn to let go of whatever he has in his mouth for something better when he hears the cue "Drop-it." After a few weeks, the puppy will automatically drop what is in his mouth on cue because he has been conditioned to do so.

While in the PBS, demonstrate what training the cue *drop-it* or *leave-it* looks like, then ask the client to work with her puppy to make sure she is going in the right direction to train the behavior. Before ending the training session, show the client how to take items out of the puppy's mouth safely. Let the client demonstrate to you how she can take an item from the puppy's mouth.

Depending on which cue you are teaching the client, send home the appropriate client handout in Appendix D. If you choose the cue *leave-it*, the PuppySmarts Chewing lesson demonstrates this technique nicely.

Just Watch

The *just watch* cue is commonly used when you are working with a puppy who wants to chase after something. This can be a car, bike, child's toy, or any other moving object. Ask the puppy for a *sit* first; then gently hold the puppy in place by hooking your thumb downward in the front of the puppy's collar by the chest, with the rest of your hand resting on his chest. The easiest way to do this is by getting on the floor at the puppy's level and placing the puppy between your legs. This way, when the puppy tries to lunge or goes to chase something you can give the verbal cue "Just watch" and gently hold him in place until he settles down.

Once the puppy stops the lunge or forward movement toward the moving object, mark and reward him. Repeat this exercise a few times a day by using something the puppy would normally want to chase or lunge at until he becomes consistent at responding to the cue by settling down and just watching the moving object.

Quiet

The *quiet* cue is commonly used when you are working with a puppy who is barking excessively for one reason or another. In many cases, head collars work well when training this cue. When the puppy begins the excessive barking, gently tighten the head collar around the muzzle and give the cue "quiet" in a soft, calm voice. Do not give this cue in a loud voice or the puppy may think you are joining him in his excessive barking session. The second the puppy is quiet, release the pressure on the muzzle and mark and reward the *quiet* behavior.

Repeat this exercise every time the puppy barks excessively over a few weeks until every time the puppy hears the cue "quiet," he stops barking.

Release

The *release* cue is commonly used in conjunction with the *stay* cue and sometimes with the *come* cue. This cue tells the puppy that the requested behavior is complete. I use the words "free dog," but this can be any one- or two-word cue that you or the client chooses. For example, if you put the puppy into a *sit/stay* or *down/stay*, when you mark and reward the puppy for holding the *stay*, you use the *release* cue to tell the puppy that the *stay* is complete and he is free to move.

Sit

The client handout Sit in Appendix D will provide you with good information on training the puppy to sit on cue. In most cases, sit is an easy cue to teach. Below are some ideas on working with puppies who seem to have a challenge with the *sit* cue.

Some puppies resist the suggested method in the client handout or on the PuppySmarts training videos by backing up. You can address this resistance in several ways. You can ask for the cue in a corner so that the puppy is prevented from backing up, or you can try asking for the *sit* on a different surface.

Some puppies are more willing to *sit* if you place them on a rug or pillow as a different texture for them to sit on. This can make a difference when the cue is first being taught. Once the puppy is consistent on sitting on the special surface, you can then begin to train the *sit* on different surfaces.

One area that can confuse patients and clients is when to mark the behavior and offer the reward. A client may ask her puppy for a *sit*. The puppy will sit briefly then pop up, and the client ends up giving the puppy the mark and reward when the puppy is actually standing instead of sitting. This confuses the puppy, as he will relate the mark and reward to standing, not the *sit*. To see whether this is what is going on with your client, ask the client to demonstrate how she is training the puppy to *sit*. In many cases, the client will re-create the same mistake in front of you, thinking she is doing the training correctly. This is your opportunity to first acknowledge what the client did and then explain what just happened.

You can then demonstrate how *sit* can be taught, marked, and rewarded in a way that gives the animal a clear understanding of what you are asking for. When you ask the puppy for a *sit*, you will need to quickly give the mark. As you reach down to give the puppy the reward, in most cases, since the puppy has been trained to stand for the reward, he will try this with you. The second his rear comes off the floor, pull the treat away from him. Most puppies will again *sit*, wanting the treat. When he offers the *sit* again, mark quickly and reach down to reward him. Keep pulling your hand with the treat in it away from the puppy until he sits, and then quickly offer him the treat. Once you get the puppy to do it correctly, demonstrate it one more time for the client and then ask her to repeat what you just did. Many clients as well as puppies will need a little coaching, but they will understand rather quickly.

If you are still having problems getting the puppy to *sit* on cue, consider using a different surface. This could be a rug, dog bed, or pillow. Pay close attention to the puppy's body language. Some puppies will give you clues on where they are comfortable and where you can get them to *sit* on cue.

As an example, a number of years ago, I was soliciting a practice to send their patients to me for puppy classes as well as private training sessions. The owner of the practice told me she had many trainers come in who wanted a referral from her practice, yet not one of them could get her dog to *sit* on cue. I told her I would be happy to give it a try. We made an appointment for a few days later when she would bring her dog into the practice.

The day of the appointment, her little chihuahua was running around her office and then back to his bed under the doctor's desk. Once in his bed, he would sit there for a few seconds then jump up on the veterinarian's leg. I asked the doctor if I could bring the puppy's bed out from under her desk, which she agreed to. I lured the puppy over to the bed and asked for the *sit* once the puppy was on his bed. He sat immediately for the treat. The veterinarian could not believe he did it so quickly and told me it was a fluke and to do it again. After the third or fourth time the puppy sat on the bed, the veterinarian asked, "Does this mean the only place he will sit is on his bed?" I replied with, "Let's see."

I lured the puppy off the bed and asked for a *sit* on the tile in her office. This time, I stood there and waited 20 to 30 seconds for the puppy to *sit*. By this time, I was praying that he would sit and envisioning him sitting in front of me. Finally, he sat, and I gave him a jackpot of treats. The doctor was convinced I knew what I was doing, and her practice became one of my clients.

The only thing I did differently than the other trainers probably was to pay attention to the puppy. He showed me where he was comfortable sitting. All I had to do was follow his lead.

If you find the puppy seems to be resisting the *sit* cue, and you have tried different surfaces, do not push the puppy's bottom down to the floor. There may be a medical reason for this resistance (e.g., hip dysplasia or knee problems, among others). Instead, recommend that the client make an appointment with her veterinarian for a complete puppy checkup. I have worked with some puppies that showed concern with the *sit* cue and later on were diagnosed with hip problems. This is not always the case, but it is better to be safe than sorry.

Stay

Read the client handout on Stay in Appendix D prior to your PBS. This overview reviews a few mistakes clients make when training this cue. When training the cue

stay, the training should begin in a quiet location with little or no distractions. Start off by asking for a three-second *stay*. Once the puppy has remained in the exact same location for three seconds, mark the behavior softly with "Yes" or a click from the clicker, and reward the puppy while he is still in the *stay*. Quickly give the puppy the *release* cue.

A *release* cue is a must when training this behavior. I use the words "Free dog" as my *release* cue; however, the client and/or you may choose any word that you like, but the same word must be consistent throughout the dog's life.

When training the *stay*, the client and you must gradually work up the length of time the puppy is asked to *stay*. If the puppy breaks the *stay* prior to the *release*, then you are moving too quickly. Remember to always train at the animal's pace. Once the puppy is consistent with the *sit/stay* or *down/stay* in a few different areas, distractions can be added. Once the puppy is consistent with giving the requested response to the cue with distractions, the training can be done at the front door or outside.

Watch for the following problems in teaching the *stay*:

- *Unrealistic expectations.* Expecting a puppy to sit for 30 seconds to a minute in the beginning is just not realistic. Many clients expect their puppies to *stay* for an extended period of time, and all too often the puppy will break the cue. Clients need to understand it is very important in training that they build on their puppy's success. It is better if the client asks for a one- to three-second *stay* in the beginning, and then quietly mark the *stay*, reward, and then *release* the puppy. Once the puppy can *stay* without breaking the cue, additional time can be added. Increases in length of time should be gradual; for example, 1-3 seconds, 3-5 seconds, 10 seconds, 15 seconds, 20 seconds, 30 seconds, and so on.

- *Inappropriate marking or rewarding.* When the client gives the mark or reward when the puppy breaks the *stay* command, the lesson is lost. Marks and rewards should be given only when the puppy has stayed in the exact place in which he was originally asked to give the behavior.

- *Breaking the* stay *cue.* Remember, puppies are constantly learning. If the client asks for the *stay* in one place and the puppy breaks the *stay* by moving up a few inches, and then the client asks for the *stay* again in the new location the puppy moved to, in time the puppy will learn he can adjust the *stay* cue to whatever location he wants. "Stay" means "Stay exactly where the puppy was given the cue the first time." If the puppy breaks the *stay* cue, then the client must gently put the puppy in the exact same place the behavior was originally requested, not a few inches in a different direction. Otherwise, in a very short time, he will begin scooting up from where he was placed to where he wants to go and the cue *stay* will lose its effectiveness. The best way to get the puppy back to the original location of the *stay* is by using either targeting or a lure.

It is important to keep eye contact with the puppy. In addition, the client needs to reinforce the word "Stay" when training this behavior.

As mentioned previously, always use a *release* cue. Every time the puppy completes the *stay*, the client must mark, reward, and then *release* him from the *stay*. To see what this training looks like, review the PuppySmarts Obedience video. This may also be a good training tool for the veterinarian to prescribe to the client at the end of your PBS.

Target

To begin target training for larger breeds of dogs, use the palm of your hand as a target. Place the palm of your hand in front of the puppy's nose. Say the word "Target," and wait for your puppy to lean forward and touch your hand. The second your puppy leans forward and touches the palm of your open hand with his/her nose, mark the behavior (the puppy touched your hand) and give him/her a treat to reinforce the behavior the puppy just gave you.

With smaller-breed puppies and kittens, you can use a targeting stick or an old wooden spoon. Take a piece of colored tape or use a marker and draw a line around the spot on the stick or spoon you want the pet to touch. Place the stick or spoon an inch in front of the animal's nose. The second the puppy/kitten touches the targeting stick or spoon, mark and reward your pet.

If your pet does not reach out to touch the object, try rubbing a little cheese, chicken, or fish on your hand or on the targeting stick. This will usually get the animal's attention. Repeat the above exercise with the new smell on the object.

Once your animal consistently touches the object when you say the word "Target," extend the distance between the animal and the target or your hand. Remember, every time you ask the pet to *target*, it is your job to mark and reward when he/she demonstrates the desired behavior.

Continue to gradually extend the distance until the puppy/kitten will walk around in a circle or across the room to *target* the targeting stick or your hand when asked to do so.

Watch

This cue is commonly used when you are working with a dog that shows leash aggression toward other dogs. Leash aggression is not addressed in this book as it is not a common puppy behavior.

When a dog shows leash aggression toward other dogs you want the dog to associate something good happening every time he sees another dog. This can be accomplished by training the cue *watch*.

Start off in a quiet place with little or no distractions and say the word "Watch." The second the dog looks up at you, mark and reward the dog. Continue training this cue until every time you say the word "Watch" the dog consistently and immediately looks up at you. Once that has been accomplished you can start adding other dogs as distractions.

At first you will want to be on the opposite side of the street of the other dog. As soon as you see the other dog, start asking for the *watch* cue to keep the dog focused on you and not the other dog. As the training for this behavior moves forward over the next few days, you can begin to shorten the distance between the aggressive dog and other dogs. Always keep yourself between the two dogs until the aggressive dog can walk past another dog politely. Once you have completed this counter-conditioning, which can take a few weeks to do, many leash-aggressive dogs will actually enjoy greeting other dogs.

This can also be practiced at a dog park if the puppy has completed his shot series and received his rabies vaccine. In this environment, the client can let the puppy play with other dogs and, after a while, put the puppy on leash and practice the *watch* cue while at the park. Every time another dog approaches while the puppy is on leash, the cue "Watch" is given to keep his attention on the client and not on the other dog. When the puppy looks at the client, the client marks and rewards the behavior.

Note: The reason you use the word "Watch" instead of "Just watch" for leash aggression is because you want an immediate response to the cue. Using additional words when working with leash aggression can cause a slower response from the animal. "Watch" essentially means "Look at me" and "Just watch" means to settle down and watch instead of chasing something.

Puppy Behavior Challenges

Aggressive/Bully Puppies

Some aggressive behavior may be normal for a puppy, depending on the situation. Aggression without provocation is another matter. There are several reasons puppies display inappropriate aggression. Many puppies labeled as bullies or as aggressive are actually fearful puppies. Others may have medical reasons for this behavior and/or medication may be needed. Make sure these puppies are cleared by the veterinarian prior to a PBS and that the doctor wants you to address this behavior and has noted it on the patient's chart.

Aggressive/bully or maybe fearful puppies require clear guidelines, confidence, and a lot of controlled socialization. They need to be desensitized to certain situations, people, or other things. They need to learn that their new family members control all resources and are the givers of all good things in life. That means the client controls when the puppy eats, where the puppy sleeps, and when the puppy is petted, played with, or exercised. None of these decisions should be made by the puppy. Puppies do not grow out of bad habits; bad habits just get worse and escalate if they are not addressed properly. The younger the puppy is when the behavior is addressed, the better the outcome can be. Consider using touches on these puppies. Touch information is located in the section on Alternative Methods in Chapter 3.

If at any time while working with a puppy, you or the client feels threatened or concerned about safety, stop the session immediately. A referral to a behavior professional is in order. Follow your practice's protocols and excuse yourself from the room. Share your concerns with one of the veterinarians. It may be appropriate for the doctor to come in and make the referral to a behavior professional. The PBS should be ended by the doctor.

In some cases, what may look like aggression can be a fear-based behavior. If the puppy is younger than 13 weeks of age, the puppy may be getting his strength from the caregiver. When separated from the client, you may see a very different behavior present itself. The puppy may cower and show signs of being afraid. Socialization, desensitizing, and not carrying the puppy may prove helpful. Puppy classes would be very beneficial for these puppies. If classes are not offered in your area, the client can begin socializing and desensitizing the puppy with your guidance (refer to the socialization and desensitization entries in the Training Terminology section of Chapter 3). The information in the client handout on Socialization in Appendix D will be helpful as well. Always stay focused on the task at hand when working with these puppies in the PBS as well as on the examination table. Safety should always be your first concern.

Review the client handout on Aggressive/Bully Puppies in Appendix D for additional information on addressing this behavior. There is no cure for true aggression, nor are there yet any FDA-approved drugs the veterinarian can prescribe that can stop the problem. Living with an aggressive animal requires a lifelong commitment from the owners. The age of the puppy when the aggressive behavior begins is important. Composition of the family structure will play an important role in the direction and decisions the client makes about the pet (young children are always going to be a higher risk to consider). Clients may need to understand and make some hard decisions if they want to keep an aggressive puppy in the family. Before decisions like this can be made, the client should be referred to a behavior professional and the referral should be noted and initialed in the patient's chart.

Many puppies younger than 13 weeks of age can be helped. The client must have clear guidelines for the puppy to follow, control all of the puppy's resources, and make all the decisions on when the puppy is played with, when it is petted, and where the puppy will sleep. If the puppy is blocking the client's passage through the home, then the puppy should be asked to move. I use the cue "Excuse me" when training a puppy to move. The client can gently touch the puppy's side with her foot and ask in a gentle voice, "Excuse me." When the puppy moves, the client should mark and reward the desired behavior with a treat, ear scratch, or belly rub, whichever the puppy enjoys most. If the puppy growls when touched on the side, the client must ignore the growl and stand there and wait, if possible, for the puppy to move. It is important that the client does not try to stare down the puppy. Staring the puppy in the eyes may be misunderstood by him that she is challenging him. Have the client stand there and wait, but look away from the puppy, if possible. When the puppy does move, the behavior should be marked and rewarded.

The following sections present a few ways in which aggressive behavior may manifest itself.

Food Aggression/Guarding

The puppy may be being teased by children or other family members while eating. Over time, the puppy may begin to growl when approached while eating. I would consider this a reasonable aggressive behavior for a puppy if this is the situation.

With puppies younger than 13 weeks old, if the teasing is stopped the behavior may stop as well. The client should make sure her puppy gets to eat in peace for at least a week. Then the client can begin offering small portions of the puppy's meal, so the puppy can eat the few pieces in his bowl and will then look up to see where the rest of his food is. When he does, the client can add a bit more at a time until the puppy has finished his normal amount of food. During the following meal, the client should first scratch the side of the puppy's neck, tell him what a good boy he is, and then lower the food bowl to the floor with just five or six pieces of kibble in it. If the puppy does not growl at the client while eating his few morsels, then the client should stay with the puppy and gently continue scratching his neck occasionally while he is eating. If the puppy growls, have the client simply walk away from the puppy. The client can repeat the exercise in two or three minutes.

If the puppy again growls, then the client should again walk away. Inform the client that if the puppy growls at this point, she should contact your office for a referral to a veterinary behaviorist. After 10 to 15 minutes, the client can confine the puppy to his crate, put the food bowl down and then let him out of his crate. Have the client walk out of the room and ignore the puppy. When the puppy is done eating, the client can take him out to relieve himself, but should give little attention to the puppy other than to mark and reward the puppy for eliminating in the proper area. Any attention the puppy gets must be initiated by the client and not in response to the puppy's demands. In many cases, these puppies can also be demanding in other situations with the family. This is when consistency and clear guidelines must be set. All good things in life happen to the puppy when the client makes the decision to do so.

Toy Guarding

Again, you will want to find out if any teasing is going on with the puppy by any family members. If so, the culprit must be stopped. If the puppy is guarding one specific toy, the toy should be thrown away when the puppy is not playing with it. Toys can then be introduced slowly, depending how the puppy responds. If the puppy growls when playing with any toys, the client may need a referral to a veterinary behaviorist.

Physical Aggression toward Humans

The first step when physical aggression toward humans is a problem is a complete physical examination by a veterinarian. The aggression may be an indication that the puppy is in pain. If the puppy is found to be in good health and the doctor cannot find a medical problem that may be causing this behavior, a referral to a behavior professional is in order.

Physical Aggression toward Other Dogs

If the puppy is willing to cause physical damage to another animal in the home, especially around 5 to 18 months of age, refer the client to a behavior professional.

If the puppy seems aggressive with other dogs only while on leash, then the cue *watch* can be trained. See the section on the cue *watch* in this chapter. This cue is trained very similar to *pay attention*. Instead of using the puppy's name as the cue, use the cue "Watch." Similarly, every time the client takes the puppy for a walk and another dog approaches, the client can give the cue "Watch," and then mark and reward the puppy with a treat for looking at her instead of the other dog. Have the client start off by crossing the street when another animal is approaching. Every time the puppy's attention strays to the other dog, the cue "Watch" is repeated. Every time the puppy turns his attention to the client, the behavior is marked and rewarded. This will distract the puppy, and over time the puppy will become conditioned that every time he sees another dog, good things happen when he looks up at his puppy parent.

Many cues and rewards in succession will be needed in the beginning to keep the puppy's mind off the other dog across the street. After a few weeks of practicing, the client can wait a little longer each time before crossing the street while keeping the puppy's attention on her and not the other dog. In time, she will be able to walk on the same side of the street as the other dog, but she should always stay between her dog and the other dog to deter lunging.

One method that has worked well in my puppy classes whenever a puppy barked and lunged at other puppies would be to ask the student and puppy to come into the center of the classroom while everyone else took their seats. When the puppy and student were in the center of the room and the puppy began to lunge or bark at the other puppies, I would ask the student to throw the leash on the floor and walk out of class and to close the door behind her. Once the puppy realized he was all alone in the center of the room, a few different things would happen. Some puppies would look for a corner to run into and cower fearfully, some would sit in the center of the room and urinate submissively, and others would freeze in place. Once the puppy showed any of these behaviors, I would invite the student back into the room and ask her to greet the puppy as if nothing had happened. If the puppy started to lunge or bark again, I would repeat the exercise. Usually within three tries, the puppy would settle down and the barking and lunging would stop. Very seldom would this exercise need to be repeated with the puppy again.

So if a puppy is lunging or barking at you during a PBS, ask the client to drop the leash and walk out of the room and see what the puppy does. If the puppy freezes, urinates, or tries to hide, then socialization and desensitizing the puppy as described above will prove very helpful. (*Note:* Try the above-mentioned method in your PBS only if the puppy is younger than 14 weeks old. If the puppy is older and you feel comfortable with him, then give it a try. If you do not feel comfortable with the puppy, though, the patient's veterinarian may need to consider a referral to a behavior professional.)

If the puppy is a small breed, the client may be carrying the puppy around too much and may need to be instructed to let the puppy explore the world with four paws on the floor. Sometimes the client is afraid of other dogs, especially large dogs. If this is the case, the puppy may be picking up the client's concerns and acting accordingly. Dogs do pick up on our emotions.

Barking

Basic alert barking is normal for dogs. Excessive barking is another thing. Excessive barking can mean a lack of socialization, exercise, or perhaps the need for desensitization. When you speak to the client, find out what triggers this excessive barking behavior: how long the puppy barks, when it happens, how often, and whether it is only with certain situations or with many. You will also want to understand what the client considers excessive. Dogs bark—it is natural that they do. However, the puppy may be barking for several reasons:

- The puppy is exhibiting age-related vocalization if he is 7 to 14 weeks old.
- The puppy is fearful, resulting in excessive barking.
- The puppy is bored due to lack of exercise and/or mental stimulation, or a lack of socialization.
- The puppy is afraid of many things and may need work on socialization and desensitization.
- The puppy is offering a normal alert to his family that something is different.

Depending on the situation and the client, consider using touches and T-shirts as explained in the section on Alternative Methods in Chapter 3. Make sure the animal is getting his share of proper socialization and exercise.

Among the ways to address excessive barking are the following :

- When the puppy is barking, have the client leave the room.
- If the puppy is barking at something outside, the client can cover the windows.
- If the client is carrying the puppy around, have her stop carrying the puppy and let the puppy experience life with all four feet on the floor. This may stop the problem.
- Have the client use a bottle or can with coins or rocks in it. One hard shake will make a loud noise and break the barking pattern. The client should immediately give the cue "Quiet," and the second the puppy is quiet, mark and reward the *quiet* with a treat.

- Similarly, have the client use a loud whistle, like a gym coach might use. This will interrupt the behavior, and the client should immediately give the cue "Quiet." The second the puppy is quiet, the client should mark and reward the *quiet*.

- If all else fails, recommend using a small air horn. That usually stops the behavior pretty quickly, but this method may hurt the puppy's ears, so this should be used only as the last resort when all else fails. Have the client use a short toot from the horn to interrupt the behavior, as above, and give the cue "Quiet." The second the puppy is quiet, the client should mark and reward the *quiet*.

Note: As seen in the last three methods, the client can use almost any noise that will reliably interrupt the puppy's barking. The intent is to interrupt but not frighten the puppy. If the client frightens the puppy, he may develop into a noise-phobic dog.

In addition to the above methods, you can have the veterinarian prescribe a head collar for the client to work with this behavior. Some puppies accept the head collar quickly without a struggle, and others may need to be desensitized to it before you can use it to work on this behavior. Information on how to desensitize the puppy to the head collar can be found in Leash Pulling later in this chapter.

When working with barking and head collars, ask the client to do something that will re-create the behavior. When the puppy starts to bark, allow a few barks and then say "Quiet" in a soft voice. Gently tighten the leash, tightening the muzzle. When the puppy stops barking, say, in a soft quiet voice, "Yes," or give a click from a clicker and release the tension on the muzzle at the same time. Remember to reward the quiet behavior.

Then explain the steps to the client and ask whether she is ready to give it a try. Have the client re-create what will make the puppy bark. Allow the client a few tries to be sure she understands how to work with the head collar and the behavior. Prior to ending your behavior session excuse yourself from the room and have a doctor come in to prescribe the head collar for the patient. Make sure the head collar is adjusted properly for the puppy. (*Note*: The client needs to check the fit on any collar frequently, as puppies grow quickly.)

Remember, barking is one way that dogs naturally communicate with one another. It may be that the puppy is concerned about something and alerting his family. In this case, acknowledge the puppy. Say phrases such as "Thank you," "Good job," or "That'll do," or just give him a pat on the head. The puppy may feel he has done his job and will quiet down on his own.

Biting and Nipping

Biting is a normal behavior for puppies. When puppies join their new families, they must be taught that human skin is much more sensitive than theirs. Puppies need to learn that biting

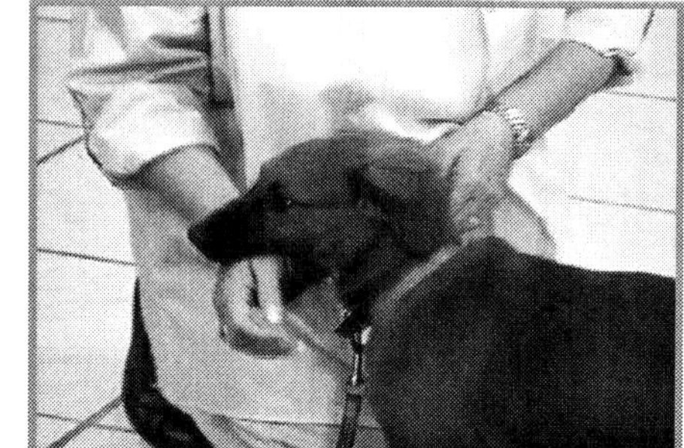

humans is not an acceptable behavior. Puppy teeth hurt our delicate skin, and their teeth are like miniature razor blades.

Old techniques such as grabbing the puppy's muzzle, giving him a shake, yelling "No," or pinning the puppy down have been shown to possibly make matters worse! These types of reprimands can be interpreted as an act of aggression by the puppy. In many cases, reprimands can escalate the problem and turn a normal behavior into an aggressive one.

When a client is playing with the puppy, and the puppy accidentally bites, the game stops immediately and the client should walk away. If the client continues to play with the puppy, the puppy receives a mixed message that biting is sometimes okay. Biting is rarely accidental, though, even in play. What you want clients to realize is that when a puppy bites, all attention must stop!

All too often, a client will make excuses for the puppy's behavior ("The puppy did not mean it," "It was an accident," "It was really my fault") and yet complain that the puppy is biting her. Clients have a million excuses for their puppies, and you need to know that is okay. When working with clients, it is easier to acknowledge their thoughts and feelings first and then let them know why biting is never okay. Explain that next time, it might be a child or it may be a harder bite. This behavior must be addressed now—it will not go away on its own and may worsen. See the client handout on Biting and Nipping in Appendix D.

When the puppy starts to bite, the client should walk away from the puppy. If the puppy is biting at shoes or pant legs, the puppy should be taken into a small room, such as a bathroom, to reenact the behavior. In your PBS, you can demonstrate this by asking the client to leave the puppy consultation room with you when the puppy tries to bite or grab clothing. Before you leave the room, explain to the client what you are going to do. Then re-create the behavior and both you and the client should walk out of the room and close the door the second the puppy bites or grabs at clothing. Leave the puppy isolated in the room. Go back in 15 seconds, make friends with the puppy, and try again to re-create the behavior. If the puppy begins to grab someone's shoes or pant leg, or bites again, walk out of the room and close the door, but this time wait 20 seconds before going back into the room. After repeating this a few times, the puppy will begin to realize that every time he bites, he gets left all alone. When clients practice this at home, be sure to tell them to puppy-proof the room they will be using to isolate the puppy.

The client will have an opportunity to see exactly what training these behaviors should look like in the PBS. Even if the puppy stopped biting after repeating the training exercise a few times in the session, that does not mean the puppy will now stop biting or grabbing altogether. This training method must be repeated a few times a day over many days, every time the puppy tries to bite. Through repetition and consistency over time, the puppy will become conditioned to the fact that every time he bites at clothes, arms, shoes, pant legs, and body parts, he will be left all alone and will not get the attention he wants.

Another way to work with biting puppies is through touches. You can demonstrate the touches by moving your index finger or index and middle fingers around the outside of the puppy's muzzle, depending on which feels more comfortable to you. If the puppy is not fighting you, you can put a little water on your index finger and gently do little circular touches on the inside of the puppy's mouth on the gums. If the puppy is comfortable with that, you can also do touches on the puppy's tongue. If the puppy tries to bite or even lick at your hand while you are working with him, stop what you are doing. If the puppy responds positively to these touches, then show the client how to do them and let her practice on your arm to make sure she is not pushing too hard and that she is actually moving the skin on your arm when she does a circular movement. If client and patient can do this successfully, suggest that the client do touches on the puppy a few times a day for just a few minutes until the biting stops.

Yet another way clients may need to address a biting puppy is to put the puppy in his crate for a time-out. However, caution the client not to drag the puppy by his collar to put him into the crate. When a puppy is dragged by the collar, the puppy will become afraid of the client reaching for him. Dragging will result in the puppy being afraid of hands and cause the puppy to shy away or run away from the client. This technique can also instigate aggressive behaviors.

Instead, either have the client pick up the puppy, if safe, or have a short, two-foot leash attached to the puppy's collar to walk the puppy to his crate. Tell the client not to talk or look at the puppy when walking him to his crate. Let the puppy walk into the crate and close the door. Then she should let the puppy calm down. When the puppy has quieted down, the client can open the door and let him out slowly. If the puppy starts to act up or tries to bite again, have the client simply close the door to the crate and walk away. Have the client repeat this several times until she can open the door to the crate and the puppy does not try to bite. In some cases, the puppy may need 5 or 10 minutes to calm down instead of just a few seconds. Whatever it takes for the puppy to settle down is acceptable. That does not mean the puppy can be left in the crate for 10 or 12 hours, however. But some puppies can take up to an hour to quiet and settle down.

Tell clients they should not open the crate if the puppy is barking. When they do, they are teaching the puppy that they themselves can be trained! Remember, puppies are constantly learning, and if that crate door is opened while the puppy is barking, the client will teach the puppy that when he barks, the client will open the door.

Many puppies are capable of barking for a very long time. This is how I believe they wear down the clients and even professionals sometimes. But we must use tough love here if we want the puppy to learn. If the puppy carries on for 30 minutes, it is tempting to let him out for a little quiet in the house, but the client should not give in. If the client gives in, the puppy will only bark longer next time, and matters will get worse, not better. Caution your clients to always wait for that moment of silence before opening the crate door.

Many puppies do not like to be held quite as long or as often as their owners may want to hold them. This presents another area of concern regarding biting. Puppies can interpret being held as confining them. Many will start off licking the client's arm or hand, their way of politely asking to be let down. Unfortunately, most clients do not understand these signs, so the only option the puppy has to get down is to start biting. Teaching clients to understand some of the signals the puppy is sending can help stop this behavior from escalating and a bite from occurring.

Hugging, kissing, and holding our animals are human needs. In general, canines do not share this need. Let's face it—how many dogs have you seen holding, hugging, or kissing one another? It is simply not a natural behavior for them.

Many of our dogs, including my own, will *allow* the hugs and kisses, but first they need the time to understand their new family. We should not expect this acceptance of hugs and kisses from young puppies. Puppies who do not understand kisses and hugs may bite. They need time to understand us and, since they are constantly learning, they will figure it out soon enough and allow us these emotional behaviors. Give the puppies the time they need to get used to some of our little idiosyncrasies. How comfortable would you be if someone walked up to you and smelled your butt? But that is a natural behavior for dogs, just like hugging and kissing is for us.

I cannot overemphasize this important point: Never leave children alone with a dog. According to statistics on dog bites in the United States, most bite victims are male children and children younger than 13 years old. Boys between five and nine years of age are bitten more frequently than others.[6]

For example, a technician student in one of my classes told the story of a dog in her community biting an eight-year-old boy. Her city ordinance states that a dog that bites must be euthanized. The dog was brought to the practice she worked at for the euthanasia. Once the dog was dead, the technician noticed the dog had eight staples in his right ear. It would seem that this terrible dog could only handle seven staples before he could no longer take the cruelty. His reward for being left alone with this eight-year-old boy and trying to be patient and not biting him was death! Puppies and children of any age should never—ever—be left alone unsupervised.

Now, we need to look at where we go if all else fails. You may find yourself in a position in which you know someone is going to get hurt by a particular puppy. You may have a small window of opportunity to promote intervention by a professional veterinary behaviorist. These people are in our profession for a reason. They have excellent skills to draw from, so use them. First, make sure the veterinarian has given the puppy a clean bill of heath. For all you know, the puppy may be in tremendous pain. If all the lab work comes back negative and the animal is physically healthy, then the veterinarian should discuss with the family either a referral to a veterinary behaviorist or the dangers of keeping an animal that is displaying these behaviors.

A veterinary behaviorist will always be your best option. A veterinary behaviorist can also offer medication for the puppy if needed. There are also animal trainers who say they specialize in aggression. If you choose to use one of them, make absolutely sure no harmful methods are used, as this will only make matters worse.

If the client refuses to talk to a veterinary behaviorist, then recommending relinquishing the dog to a shelter or euthanasia may be necessary. This must be the doctor's call. There is no cure for true aggression. Based on the family structure and commitment to the animal, clients must be warned. Aggression in some cases can be contained if worked with properly, but the animal will have aggressive tendencies for the rest of its life. Such an animal will require a lifelong commitment from the client and family members. Families with young children are always at higher risk for aggressive bite wounds.

The conversation about the dangers of an aggressive dog should take place between the client and a veterinary behaviorist or one of the veterinarians in your practice. When the veterinarian discusses aggressive behaviors with the client, the client should be asked to sign a form that states she has been informed of the dangers of keeping this puppy. To protect the practice, this form should be designed by an attorney and signed by any client with aggressive puppies or adult dogs. Make sure the client's file is documented with the warning given and that there is a signed document stating the behaviors the practice observed in the animal, the options given, and the owner's acceptance or nonacceptance of the recommendation. Follow your practice's protocols whenever you are faced with a situation you didn't expect to arise in your behavior session.

Never try to deal with aggressive biting on your own unless you personally have had many years of working with aggression and have the approval of the practice's owner or veterinarians. Do not feel that you need to know everything to do a good job. It is not your responsibility to know all the answers. Sometimes we don't have all the answers, and that is okay. This is why referring the client to a veterinary behaviorist is the best choice. In most cases, they can really make a difference. Referring them now can make the difference between the animal keeping his life or being euthanized. The earlier we address the problem, the better chance the puppy and family will have to work with this behavior. If we are lucky, the puppy might be able to keep his home and you will know you played an important role in helping this animal and client.

Bolting Out Doors

A puppy that bolts out of an open door is at high risk of getting lost or hit by a car. These are just two of the many reasons this behavior should be addressed. Each

client should always make sure her puppy wears a collar with an identification tag. All puppies should be microchipped as an added safety measure.

The first command the puppy needs to learn is *sit*. Once the puppy is consistent with the *sit* cue, then the client can introduce the *stay* cue. It is important to keep eye contact with the puppy when training *stay*. You can read more about *sit* and *stay* in the Cues/Commands section of Puppy Training Basics in this chapter, as well as the corresponding client handouts found in Appendix D.

Once the *sit* and *stay* cues have been established, the client will need to retrain these cues at the front door. The client should put a leash on the puppy so she can

step on it or grab it quickly if the puppy tries to bolt. Instruct the client to use this sequence of steps:

1. The client puts the leash on the puppy.

2. The client places a small rug by the door for the puppy to sit on. The rug should be far enough away from the door so the door can be opened without making the puppy move and positioned so the puppy can watch who is coming in.

3. The client asks for a *sit/stay* on the rug and takes one step away from the puppy. The client then waits a few seconds and returns to the puppy. The client marks softly, rewards the puppy with a treat, and gives a release cue. The release cue is always used when asking for a *stay*. It lets the animal know he is released from the *stay* and free to move around. (I use the words "free dog" as my release cue.)

4. The client must not reward the puppy if he moves from the exact spot where the client placed him in the *sit/stay*. The client should return the puppy to the original location on the rug and give the *sit/stay* cue again. This is true anytime the puppy breaks the *sit/stay*. A puppy should leave a *sit/stay* only when released and not before. (If the puppy moves and the client simply asks for a *sit/stay* in the new location to which the puppy moved, the puppy will be confused as to what exactly *sit/stay* means.)

5. Once the puppy is doing Step 3 consistently, the client can then take additional steps, one at a time, away from the puppy toward the door. The client should return to the puppy after each additional step, mark, reward, and release him until the client can put her hand on the doorknob and the puppy will not break the *sit/stay*. You want the client and puppy to succeed every step along the way. Remind the client to always walk back to the puppy, mark, reward, and release him for holding the *sit/stay*. A jackpot reward for a job well done would be appropriate here.

6. Once the puppy is doing Step 5 consistently, the client can step on the end of the leash and open the door just a little with the puppy in his *sit/stay* on the rug. If he has difficulty or will not hold the *sit/stay*, have the client back up to Step 5 for a while. If the puppy holds the *sit/stay*, the client should close the door, walk back to the puppy, and mark, reward, and release accordingly. The client should continue to increase the distance she opens the door until the door is wide open and the puppy holds the *sit/stay*.

7. Once the puppy is doing Step 6 consistently, the client can step outside the open door for a few seconds, come back inside, and walk over to the puppy to mark, reward, and release him if he has not broken the *sit/stay*. If he has difficulty or does not hold the *sit/stay*, the client should back up to Step 6 for a while. If the puppy holds the *sit/stay*, the client should close the door, walk back to him, mark, reward, and release him. Have the client increase how long she stands outside while the puppy holds the *sit/stay* until the puppy can easily hold the position there for a minute or so.

8.   Once the puppy is performing Step 7 consistently, the client should invite a friend or family member over to ring the doorbell or knock on the door. After hearing the doorbell ring, or the knock at the door, the client can then ask her puppy for a *sit/stay* on the rug with his leash on. Keeping eye contact with the puppy, the client should step on the leash before opening the door slowly. If the puppy does not hold the *sit/stay*, the client should close the door and try again. The client should put the puppy back into a *sit/stay* in the original position and again try to open the door while keeping eye contact. This may take a number of tries for the puppy to be successful. If the puppy holds the *sit/stay*, the client should open the door, let the guest in, and close the door behind her, then walk back to the puppy, mark, reward with a jackpot of treats, and release him. (Clients may want to consider offering their friends or family members that are helping with this behavior a treat too. After all, they may get the door closed in their faces a few times before the puppy is successful.) Instruct the client to have her friend ignore the puppy and not make eye contact. You want the puppy to keep eye contact with the client until he is released.

9.   The client should repeat Step 8 with lots of people in the same day. This will quickly desensitize the puppy to hearing the doorbell ring or a knock at the door and the door opening. The puppy is also learning to come to the front door to watch when someone comes in.

Over time, the puppy will learn to sit on the rug when the doorbell rings or someone knocks on the door. Have the client repeat these steps as necessary until the puppy runs to his rug and sits when the doorbell rings or someone knocks.

This process can take some time, so remind clients to always end each training session on a positive note. Also, clients will want to keep the training sessions short so they will not get frustrated or lose the puppy's attention.

This seems like a very long and involved process, and it can be. However, many puppies and dogs get this behavior in a few days. When talking to clients, always estimate a longer time for training any behavior than what it usually takes. Some animals learn quickly and others take a bit longer. It is good for a client to believe she has a very intelligent puppy or that she is a great trainer because her puppy learned so quickly.

Review the client handout for Bolting Out Doors in Appendix D. It is not as detailed as the steps outlined here because you don't want to overwhelm the client. In addition, many puppies that previously bolt out the door will learn this new behavior quickly if the client is consistent.

In your PBS, you can demonstrate training the *sit* and *stay* cues and what the mark, reward, and *release* should look like. Remember to ask clients to repeat what you demonstrated so you can make sure they will train the behavior properly.

Chasing

When puppies see bicycles, scooters, cars, or remote control (RC) cars for the first time, they may bark, try to chase them, or just watch them. A puppy that can just

watch moving objects is perfect and would not need this training. Let us explore ways to deal with puppies—especially herding breeds—that bark or want to chase moving objects. I will share a few techniques for both issues (barking and chasing) that have worked well for me.

The client may need some help from another person to drive an RC car, ride a bike, or roll a ball or toy across the floor while the client works with this behavior. If the client does not have anyone to help train this behavior, she will need to get a little creative. She can go to a park where children are running around and playing or to a safe location where people are riding their bikes or scooters.

First, I ask clients to get down on the floor at the dog's level. I suggest they put the puppy between their legs; have them take one hand and put their thumb through the front of the collar and hold the puppy's chest with the palm of their hand. Lightly hold the puppy with the other hand to keep the puppy gently confined. I have the client ask the puppy for a *sit*. Then for the chasers as well as the barkers, I give each a cue to use. I like to use the words "Just watch" with chasers and "Quiet" with the barkers.

While another person keeps objects moving, the client keeps both hands lightly on the puppy so the puppy can be controlled in a gentle way and be kept in a *sit* position. The client can use a low, calming voice, saying words such as "Just watch," "Quiet," or "Shhhhhh." The second the puppy stops trying to escape to chase the object or stops barking, the client quickly marks with "Yes" or a click from a clicker, and rewards with a treat. After eight to ten tries, the puppy will begin to understand what "Quiet" or "Just watch" means.

Once the puppy can sit quietly for 30 seconds without lunging or barking at a moving object, have the client stand up and step on the leash about three feet away from the clip on the collar. When the client and puppy are in position, introduce the moving object. If the puppy goes to lunge or starts to bark, have the client ask for a *sit* and a *just watch* or a *sit* and *quiet*. If the puppy holds the position, the client marks and rewards the puppy. This may take several tries for the puppy, as we have changed the position of the client when training the behavior. After a while, the puppy will learn to be quiet or just watch without lunging or barking at moving objects.

Note: If the client cannot get down on the floor, you can begin training the exercise from a standing position or have the client put the puppy on a table instead of the floor to work with this behavior.

Chewing

Chewing is a normal behavior for puppies and adult dogs. It is our job to make sure they know what is appropriate for them to chew on. You will want to make sure clients have safe toys for their puppies to chew. Plush toys and most squeaky toys are not much help in relieving the discomfort of cutting new teeth, nor will they satisfy the puppy's (or adult dog's) need to chew.

It is important that clients understand what kind of safe toys will satisfy their puppy's need to chew. Strong rubber toys that you can put treats in are perfect chew toys and mentally stimulating at the same time. The veterinarian should prescribe appropriate toys for a puppy to chew on prior to ending your PBS. In fact, the practice should carry any training tools you discuss in your PBS, as we live in a society where one-stop shopping is the most convenient for all of us.

The old adage is: If you do not want the puppy to chew on shoes, do not give the puppy an old shoe to chew on. The same holds true for all other items in the house. The puppy cannot distinguish between old and new or expensive and inexpensive, so make sure the clients understand this.

Some ways to work with puppies that have inappropriate items in their mouth include the following:

- Clients should first puppy-proof their homes. Puppies are inquisitive and will want to put anything and everything they can find in their mouths.

- Offer a favorite toy in exchange for an object the puppy has in his mouth.

- Train the cue *leave-it* or *drop-it* to tell the puppy to leave something alone or to drop something the puppy already has in his mouth. These cues are discussed under Cues/Commands in the Puppy Training Basics section of this chapter.

Training the cue *leave-it* is nicely demonstrated in the PuppySmarts video lesson on Chewing. This may be an excellent training tool for the doctor to prescribe for the patient at the end of the PBS.

Another option clients may use to discourage chewing is spraying objects the puppy wants to chew with something that tastes or smells bad. Suggestions include oil of citronella sprays, water mixed with a small amount of cayenne pepper, or bitter apple sprays. The client can try any of these items to deter the puppy from chewing on inappropriate items. These products are helpful for garbage-can raiders, chair-leg nibblers, and sometimes countertop surfers. You can also use a Scraminal®, as explained in the section below on Counter Surfing.

A puppy between six and nine months of age will go through a strong chewing stage as the adult teeth erupt. Many times, this behavior is labeled as destructive chewing or separation anxiety because the puppy seems to almost chew up the house when left alone.

If the puppy is simply chewing to set his teeth into his jaw, then giving the puppy hard rubber toys to chew on, especially those you can put treats into, can help. If the puppy must be left alone, a crate should be used when going through this stage, even if the puppy has been trustworthy when left alone in the past. When the adult teeth are erupting, a puppy will chew anything he can. That means *anything* left around for a puppy to chew on is fair game.

Putting a puppy in a room during this stage in his life will only lead to chewed up doors, wallboards, carpeting, and walls. A crate is the best solution when the puppy cannot be supervised.

There may be a medical reason for excessive chewing as well. Follow the protocols your veterinarians have set up for you, or consult with the client's veterinarian about your concerns. In some cases, the puppy may be having difficulty with a tooth coming in properly or a retained deciduous tooth. Either way, the puppy may be in pain and trying to deal with the problem by chewing anything and everything to fix the problem on his own.

Counter Surfing, Table Jumping, and Garbage Raiding

Counter surfing, jumping on tables, or raiding the garbage can best be addressed with the use of products that correct the behavior instead of the client addressing the behavior. If the correction occurs without the owner around, the puppy will learn to stop the behavior while young. If the correction occurs only when the client is present, the puppy will simply learn to wait until the client is gone or not watching him.

The first thing clients should do is to puppy-proof their home. If there are no interesting things on the countertops or tables, your puppy will probably not want to jump on them in the first place. However, that being said, there will be times when clients do have things on their countertop and coffee tables and the puppy will need to learn to leave those things alone.

Booby traps are an excellent way to condition a puppy that there are consequences to him when he is counter surfing. He will associate the noise and clatter of the booby trap with his action of taking something from the counter or other objects he should not be touching.

Setting the Trap

To set up a booby trap, use three empty cans as the base. Place the cans as far back on the countertop as you can. Attach a piece of string or ribbon to the front can, as shown in Figure 1.

Place a piece of cardboard over the cans. Then stack three more empty cans on top of the cardboard, as shown in Figure 2. Place another piece of cardboard over the middle stack of cans. Then stack two more empty cans on top of the cardboard, as demonstrated in Figure 3.

Fig. 1

Tell clients to leave this stack of cans on the back of the countertop for a few days or in the middle of the coffee table for the puppy to get used to seeing it. After the fourth day, begin to bring the cans a little more forward on the countertop or closer to the edge of the coffee table. Continue moving the cans closer to the edge daily. Once the cans are at the edge of either surface, it is time to bait the trap. Once baited, it is important that the client is not in the room when the booby trap goes off. This way the puppy will learn not to surf countertops or coffee tables even when the client is not home.

Fig. 2

Baiting the Trap

Tell clients to leave enough room between the cans and the edge of the countertop or coffee table for the puppy to be able to jump on without knocking the cans over. Take the string that is attached to the bottom can and tie a small piece of chicken (or something equally tempting) near the end, as shown in Figure 4. Once the trap is baited, leave the room.

When the puppy smells the bait he will jump up on the countertop or coffee table to find it. When he does try to grab it, all the cans will fall on top of him, as shown in Figure 5. This will startle him. He will look around to try and figure out why the sky fell on him. When he realizes no one is there but him, he will begin to associate his counter surfing with his experience. Some clients will need to repeat the booby trap a few times for their puppy to connect "the sky is falling" to the counter surfing or coffee table surfing. However, some dogs need to experience this only once.

Some dogs, however, may not be bothered by booby traps, so you can

Fig. 3

Fig. 4

Fig. 5

recommend a different training aid, a Scraminal or similar product. This device uses a motion sensor and emits a high-pitched sound designed to repel the puppy. Place the Scraminal on top of a countertop, coffee table, or garbage can. The noise will usually startle the puppy and he will walk away.

Either a booby trap or the Scraminal should stop these behaviors. Remember, however, that it is important that the client not be in the room when the booby trap goes off. This way, the puppy will learn not to surf the countertops or jump on tables even when there is no one home to watch him. For garbage can raiders, bitter apple or similar products can be sprayed on the garbage can as a deterrent.

Crate Soiling

Some puppies think nothing of soiling in their crates. This can be a real challenge for new puppy parents. This behavior is common in pet store puppies because they are not walked outside. It is important to first address whether there is a medical reason for the soiling. If the veterinarian gives the puppy a clean bill of health, then you can move forward with the client.

If the puppy was in the crate for an extended period of time, then perhaps the puppy simply could not hold it any longer. If that is the case, the client may want to find a pet sitter that can let the puppy out every few hours during the day. Or the client could consider taking the puppy to doggie day care once the puppy has received the appropriate vaccines.

Clients can quickly get discouraged when they have to constantly clean the puppy's crate. This is a time-consuming job, and it may discourage many puppy parents from keeping their puppies if this is not addressed promptly. There are a few circumstances to consider, though, that can lead to this issue.

The crate may be too large for the puppy. If the puppy can sleep in one area and relieve himself in another area of the crate, the crate is too big. This can be easily addressed by getting a smaller crate that allows the puppy only enough room to stand up, turn around, and lay down comfortably. If the client wants to keep the larger crate, then using a piece of wood, a metal separator, or something else to make the space smaller while the puppy is still small would help. Because puppies grow quickly, you will want to make sure the client adjusts the space as the puppy grows by moving the barrier when necessary.

The client can also consider adding a small litter box in the oversized crate if she so chooses. This way the puppy can use the litter box on one end of the crate and sleep in the other. This method is not optimum, however. It may eliminate the problem temporarily, but eventually the litter box may have to be taken out as the puppy grows. In addition, the puppy may develop a preference to eliminate in his crate.

You may also want to review the client handout on Housetraining Troubleshooting in Appendix D.

Grabbing

Grabbing occurs when the puppy takes something and walks or runs away with it. This becomes a game of chase for the puppy—a favorite game. Although the training for grabbing is similar to training the *leave-it* and *drop-it* cues, there are some differences.

When working with the grabbing behavior, it will be easier if the puppy is on leash so you or the client can grab quickly to stop the chase game. This behavior is better addressed when you do something to re-create it. For example, have a sandwich, food item, or whatever the puppy is pretty consistently grabbing available to "bait" the puppy. The second the puppy tries to grab the item, quickly grab the leash to stop the chase game from starting or, if possible, cover the item with your hand. The second the puppy is stopped, say the cue "Drop-it" or "Leave-it." If the puppy has not yet learned these cues, then you must first train him to drop things on cue by using one of them. If the puppy does not drop the item and the client considers the item to be dangerous to the puppy, have the client properly take the item out of the puppy's mouth and walk away from the puppy. The puppy should be ignored for a few minutes until he settles down. When he does, the client can ask for a behavior that the puppy knows well, such as *sit*, then mark and reward the puppy for the polite behavior. Review the information under Drop-it/Leave-it in this chapter to properly work with this behavior.

The grabbing behavior can also present itself when offering the puppy a treat. Instead of waiting to be given the treat, the puppy grabs it out of the hand. Many of you reading this book have probably made the mistake of offering a strange dog a treat in the practice and having a dog's tooth go into your thumbnail.

Here is how I have worked with the grabbing behavior successfully. The first thing you or your client needs is a long, hard treat, such as a Milk Bone®. Hold one end of the treat, and say the word "Easy" to the puppy while offering the puppy the treat. If the puppy lunges or tries to grab the treat, put the treat up and walk away from the puppy.

After a few minutes, offer the treat to the puppy again using the cue "Easy." If the puppy lunges or tries to grab it, put it up and walk away from the puppy. It may take a few days, but the puppy will figure out the message. Until he learns to wait for the treat to be brought to him without trying to grab it, you or your client should repeat these steps. Once the puppy stops lunging or trying to grab the treat from your hand, only then can he actually have the treat.

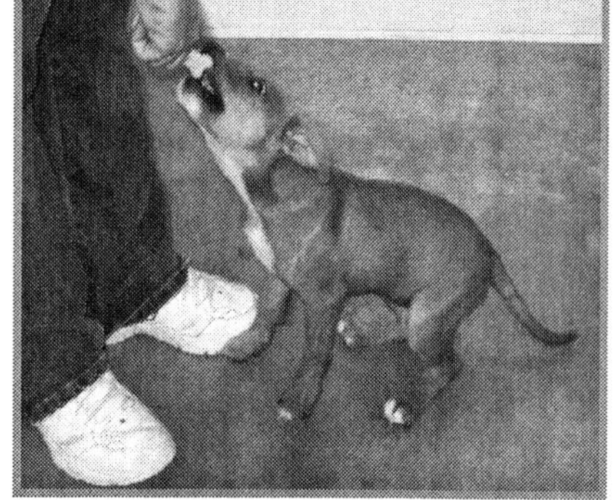

Do not work with this behavior combined with other training until the puppy is becoming much gentler. For example, when training *sit*, you must reward the puppy for offering the correct behavior. That is not the time to insist he be *easy* as well. Simply use an open hand with the treat in the palm of your hand to give the treat. If the puppy is still grabby, use the technique below to protect your fingers.

With older puppies, another way you can work with this behavior is to put the same long treat between two of your fingers with the extended treat facing outward from the back of your hand (palm down). This way, even a puppy with a large muzzle cannot bite your finger while trying to grab the treat. An open palm while offering a dog a treat is the easiest way to handle this behavior in the practice, but it does not train the animal to do anything differently, nor does it teach anything; it simply makes it safer for you at the moment.

Although dogs in general are opportunists, grabbing is a rude behavior. Many of these puppies may be used to getting their own way a bit too often. Clear guidelines are needed so the puppy understands who controls this wonderful resource.

In your PBS, demonstrate how to properly remove items from the puppy's mouth safely, as you did for the *drop-it* or *leave-it* session. Give the client an opportunity to open the puppy's mouth in your session to see that it is done correctly. If the puppy is grabbing treats from the hand, then demonstrate how to train the puppy to not grab treats from the hand. Once you show the client what to do, let her repeat what you demonstrated so she can work on this behavior at home when giving her puppy treats. This way, you will know the client understands how to work with this behavior properly without hurting the puppy or getting hurt herself. Make sure the client understands that if the puppy lunges or tries to grab the treat, the client should put the treat away and walk away from the puppy. In five or ten minutes she can repeat the exercise

Send the client home with the client handout on Grabbing or Training Leave-It in Appendix D.

Grooming

Review the client handout on Grooming in Appendix D. Discuss the appropriate grooming tools, supplies, and methods with clients before distributing the handouts. If the puppy is having serious negative reactions to being groomed after several days of using the information on the client handouts, the puppy may have a medical or underlying behavior issue that is causing this reaction. If you are concerned the puppy may have a medical reason for not wanting to be groomed, consult with a veterinarian.

When showing a client how to perform whatever grooming requirements are a problem, remember to stop the second the puppy shows concern. Strength is not how you help puppies become comfortable with grooming. When you stop, the puppy begins to trust you, and the next time you try to do the same thing, the puppy will let you go a bit further. Besides the Grooming handout from Appendix D, the client may also benefit from doing touches on her puppy and/or being introduced to Bonding Moments, discussed in Chapter 3 in the section on Alternative Methods.

Home Alone—Separation Problems

The term "separation anxiety" is a medical term. Because of this, PBAs, trainers, and even applied animal behaviorists should not use this term without a veterinarian first diagnosing the puppy with this disorder. Only a veterinarian can diagnose a puppy with separation anxiety. Sometimes the doctor may prescribe medication for this patient. You will want to make sure this is the behavior the doctor has written in the patient's chart for you to address during your PBS.

Some puppies go through a vocalization period between eight and fourteen weeks of age. The barking, crying, or whining the puppy is displaying may be normal for the particular developmental stage of the puppy. This is not uncommon. Let the client know this may be a stage the puppy will outgrow if the behavior is ignored. Caution that if the client pays attention and makes a fuss over the distressed puppy, the behavior will get worse.

Puppies that are adopted from shelters may also display this behavior. They have a good reason to be concerned when left alone—after all, look where they ended up.

Review the client handouts for Home Alone and Socialization in Appendix D as well as the definitions of socialization and desensitization in the section on Training Terminology in Chapter 3 prior to your PBSs. They will give you a better understanding of ways to address this behavior with your clients.

- *Socialization/desensitization can be a solution to this problem.* The puppy may need to be desensitized to the client leaving the house or require more socialization opportunities to build confidence.

- *Using T-shirts may also work.* T-shirts can be used with this behavior to give the puppy a better understanding of his own body. The client can put the T-shirt on the puppy about 30 minutes before leaving the house. This is an exercise that is usually easy for a client to do.

- *Using touches is another alternative.* Doing touches on these puppies can help them cope a bit better with the stress of being left alone. Using touches coupled with a confidence course and a T-shirt will help tremendously.

Give the client a few ways to work with this behavior so both client and patient will have a better chance for success. At the same time, however, you will not want to overwhelm either of them. Keep in mind you always want to build on both the client's and patient's successes. Encourage the client and let her know that many puppies do get past this period with some desensitizing, socializing, confidence-building, and time.

There are times when clients, without realizing it, may instigate this behavior. Long good-byes or conveying guilt for leaving the puppy will increase the behavior. Puppies are very good at picking up on human emotions. Make clients aware that puppies can sense when they are feeling guilty or sad about leaving them alone. When the puppies sense the clients are concerned, they will become concerned, too, and act accordingly.

Share with your clients that puppies do need a lot of sleep while young—eight to twenty hours a day. When clients do need to leave their puppies in crates, ask them to consider it nap time instead of having the thoughts of leaving their poor little puppies all alone in their crates.

Housetraining

The key to successfully housetraining any dog is to manage his environment, keep the puppy on a feeding schedule, and be consistent. When training rules are broken frequently, this causes confusion for the puppy. When puppies are confused, they cannot be successful.

Here are a few basic rules you can relay to the client:

- *Do not take the puppy for a walk to eliminate.* Instead, take the puppy to a designated place to relieve himself, and then go for a walk as a reward.
- *Bring a treat outside with the puppy.* The second the puppy is done with his business, mark and reward the behavior *where it happened.* Give the puppy the treat outside where he did his business.
- *Keep an eye on the puppy the entire time that he is out of his crate.* This means no distractions such as cooking, watching television, or talking on the phone. That is how accidents occur.
- *Place the puppy in his crate, a play pen, or a confined area* of the house when the client or other family members cannot manage his environment.
- *Watch for signals from the puppy* alerting the client to the fact he is looking for a place to eliminate.

Share with clients what some of these signals can be. These include behaviors such as a high-lifted tail, sniffing the ground, circling, or a puffed-out rear. They can all be indications that the puppy needs to relieve himself. Some puppies jump on the client's leg, others sit and look at the client, and still others sit at the door where they are normally taken out. Puppies all seem to have their own indicators, so it is the client's job to figure out what her puppy's unique indicators are.

Puppies need to relieve themselves on a fairly regular schedule, as follows:

- When they first get up in the morning
- After a nap
- 5 to 10 minutes after drinking
- 10 to 20 minutes after eating
- Before going into their crates
- When they first come out of their crates
- Before they go to bed at night

Puppies also need to relieve themselves frequently (every hour or so) when awake and active until they are about 12 weeks old. Especially during playtimes, the client or children can miss the puppy's signal to relieve himself. The puppy, getting caught up in all the fun, may wait too long to go and will simply squat or lift his leg without additional warnings.

Suggest that clients offer a special treat just for housetraining. The only time the puppy gets that particular really great treat is when he goes outside or to his designated area inside to relieve himself.

Small dogs usually take a little longer to be housetrained. They are so close to the ground that many clients do not realize when they are eliminating until it is too late. As a result, when little ones are not properly supervised, they can go potty in the house and no one notices until after the fact. The more this happens, the more comfortable the puppy gets with relieving himself on a rug or on tile. Over time, these soiled areas can become the puppy's preference as to where he wants to relieve himself, making housetraining harder.

Clients may also end up with problems if they scold the puppy for having accidents in the house. This will cause many puppies to hide when relieving themselves so their puppy parents cannot see them.

If a puppy begins to eliminate in the house, the client can interrupt the puppy by making a short, quick sound such as clapping her hands or shaking a can or plastic bottle with rocks or coins in it. In many cases, the puppy will stop eliminating and can be taken to the designated area to finish elimination. If owners scold their dogs or punish them, they will find mistakes hiding in their homes.

Housetraining a puppy requires many components for success. Let us take a look at what those requirements are and why they are important.

Managing the Puppy's Environment

This is applicable even if the client is home with the puppy but she cannot properly supervise him (e.g., when cooking dinner, helping a child with homework, or talking on the phone). In these cases, the puppy needs to be placed in his crate or confined in a small area.

Feeding must be done on a schedule. Have clients put the puppy's food down for 15 minutes. If the puppy does not finish, cover it up, and offer it to him at his next scheduled feeding time.

Clients must be sure to properly deodorize all accident sites except the area inside the house they have chosen to trigger the elimination response. One deodorizer and stain remover I have found to work well is a product called Kick®. There are many others that your clients could use but they must use something that completely removes the odor.

While sleeping, the puppy can hold it in much longer. Add one hour per month for the age of the puppy to determine how long the puppy can hold it. For example, a two-month-old puppy can hold it for three hours when asleep, and a three-month-old puppy can hold it for four hours when asleep. Puppies do not begin to control the urge to relieve themselves until about eight and a half weeks of age. Whether the client takes the puppy to paper or outside to eliminate, the mark and reward must occur when and where the desired behavior is given.

If clients did not receive the PuppySmarts Potty & Crate training video or the AAHA brochures when they first visited the practice, send them home with the clients after your session. This will help them with crate training and housetraining their puppies, making their job and yours much easier.

Common Mistakes

One common mistake is taking a puppy for a walk to relieve himself during housetraining. Being outside is wonderful for most puppies. There are so many interesting smells and sights, it is easy for puppies to get distracted. This is why it is so important to take the puppy first to a designated area, and stand there patiently waiting until the puppy has eliminated. The client should give her puppy about six feet of leash and stand in the specific area where she wants the puppy to learn to eliminate. This will allow the puppy room to sniff around the six-foot perimeter area to relieve himself.

The second the puppy finishes doing his business, mark the behavior (elimination) in a soft voice or use the clicker and follow up with a reward. The mark and reward must be given as soon as the puppy is done. Walks can be given as a reward, but never before the puppy has done his business in the designated area.

Another common mistake many clients make is taking the puppy outside to relieve himself, then coming inside to give the puppy a treat. This actually can cause additional problems. The puppy goes potty outside and wants to quickly come into the house for his reward, so he does not finish doing his business outside. Once the puppy gets his wonderful treat, he can then look for a place indoors to finish his business. Clients get discouraged, believing they just took the puppy out and now he is relieving himself again in the house. But the puppy thinks he is being rewarded for coming into the house.

The puppy cannot relate the reward to the behavior when the behavior is given one place and the reward or reinforcement is given in another. When a behavior occurs in one place and a reward given in another, it only confuses the puppy. Rewards must be given when and where the behavior occurs, not 30 seconds or a minute later in a different place. We want the puppy to learn by conditioning him that when he eliminates in the designated area, good things happen, so he will want to give more of the desired behavior.

For additional insight on housetraining challenges, review the client handout on Housetraining in Appendix D.

Crate training is a great tool in making housetraining easier on the puppy and the family. You can refer to the client handout on Crate Training in Appendix D as well.

Hyperactive Puppies

We have all seen puppies that cannot stay still for two seconds without jumping, lunging, biting, licking, or barking. The danger here is that the puppy is not learning self-control while still young. I consider this an out-of-control puppy. In many cases, out-of-control puppies are a direct result of their environment. This behavior can be the result of: (1) too much stimulus; (2) not enough guidance; (3) puppy demands attention and gets it (is rewarded/reinforced through touch or words); (4) a need for exercise; or (5) a lack of consistency. In many cases, guidelines for the puppy have not been set or are unclear to the puppy.

To stop this behavior, you need to do two things: Take all energy and attention away from the puppy when he is inappropriate, and catch the puppy being quiet so you can mark and reward him. After all, quiet is what you want more of and you need to let him know that. You should look at this puppy like you would a two-year-old child who is having a temper tantrum in the store when he or she wants something and mom says "No." If the mother walks away from the child, the child will stop the tantrum because, just like the puppy, the child does not want to be left alone. If instead mom tries to negotiate or calm the child down, the tantrum escalates and the child continues the tantrum until the mother finally gives in. Puppies and children should never win these contests. A demanding puppy becomes a demanding dog, and demanding dogs can become aggressive dogs.

If clients take all energy away from this behavior, it will diminish in the length of time displayed, intensity, and duration. The more opportunities clients have to work with this behavior, the faster it will dissipate. When working with such puppies, try to have the clients do everything on their own time schedules, not on the puppies' time schedules.

As an example, if the puppy nudges the client's arm for a pet, she should ignore him and walk away. Later on, when the puppy is lying down quietly, she can then walk over to the puppy, say, "Yes, good, quiet" in slow, soft words and offer an ear scratch. If the puppy jumps up or barks, demanding more attention, she should again ignore him and walk away.

If the puppy jumps up, it is effective for the client to cross her arms, turn her back on the puppy, and take a step forward. If the puppy jumps again, the client should put a leash on him for more control. The next time the puppy jumps up, she should step on the leash to make sure he cannot hurt her, and take all attention away from him. When the puppy quiets down, the client should calmly say, "Yes, good, quiet," and give him a reward for being appropriate. In time, the puppy will learn that he gets more attention when quiet than when he barks, jumps, or lunges, demanding attention.

Truly hyperactive, out-of-control puppies can be dangerous. When looking into these puppies' eyes, you will notice the pupils are dilated and you will have trouble seeing any part of the iris, which is the eye color. It is as if the puppy is physically there but there is nobody home upstairs; I'm sure you can understand exactly what that means. At that moment in time, the puppy is not capable of thinking. That is another reason why giving him time to calm down is so important.

Unfortunately, children and puppies are capable of outsmarting and outwaiting clients and parents. The difference is that most people will not discard their children because they are naughty, but they will discard their dogs.

You will find that many clients with hyperactive puppies are very frustrated with them. They do not know how to address the problem, and they do not want anyone hurting their puppies. Let them know you can help both the puppy and them to better understand one another without hurting the puppy in any way. In fact, you could call it a holistic approach to working with the puppy because you will be using methods the puppy naturally understands.

Training self-control is easy if you can be patient and consistent. Remember, this puppy has had a lot of experience getting what he wants. Your job in addressing this behavior is to take all attention away from him to the extent that you do not even talk about him while you are with him.

Animals seem to be able to sense when we are talking about them. When you are in the puppy consult with the hyperactive puppy and client, sit or stand quietly with your foot on the leash, giving the puppy about three feet of space between the collar and clip. If the puppy can still jump or he bites at you, then make the leash short enough so he cannot hurt you. Always ensure your own safety first and protect yourself from accidental injuries. Do not pull the leash so tight that the puppy cannot move, however; this will not teach the puppy anything. The puppy must have enough room on the leash to make the decision to sit or lie down on his own, without being forced.

Usually, the shortest leash length you should go is 18 inches from the clip to your foot, but if you must go shorter, then do so. Safety is always the first concern. Tell clients that if the puppy tries to bite them, shorten the leash so the dog cannot jump or reach any skin to bite them. It is not necessary to step on the leash near the collar as this will not teach or allow the dog to settle down properly on his own. This is not about causing pain or pinning the puppy—it is about getting the puppy to settle down. Always mark and reward a quiet puppy that has settled down. This is the behavior you want, and by marking and rewarding you are letting the puppy know this.

By keeping the puppy close, it is easier to mark and reward the controlled quiet behavior. Wait for the puppy to let out a big sigh—that says he is ready to settle down. When he is quiet, move your hand slowly down to his side and give him a long pet on the side of his body. In a low, quiet voice, say, "Yes, good, quiet" in long, exaggerated words. If the puppy starts to get up, pull all attention away from him and ignore him again. Most puppies will try three or four times before they can really sit still for a soft, quiet pet. They actually learn to settle down very quickly after the first lesson, which is usually the most time-consuming.

A few years ago, I was invited to be on a cable program that was hosted by a veterinarian. The doctor asked me if I would be willing to work with a dog on the air so his viewers could watch. I agreed. Knowing that a timid or fearful dog would not move in front of a camera, I flew to his city a day before the show. I went to the shelter to pick out eight or ten dogs that I felt could handle the studio, lights, cameras, and staff. Feeling comfortable after doing this, I went out for a lovely dinner and then went back to my hotel room.

About 10:30 that night, I got a call from the veterinarian who hosted the show, asking me whether I could work with his client's eight-month-old labradoodle on the air. The wife of his client, who was also his good friend, had threatened that if the dog did not settle down, he had to go!

When I arrived at the studio, I was sent to a small room down the hall. When I opened the door, a 65-pound labradoodle puppy lunged at me. I am happy to report he was on a leash. I stopped my forward movement and withdrew myself far enough away so he could not reach me. I asked the owner to shorten the leash to about three feet. After he did this, I decided there was no time like the present to begin working with this dog. I gathered some information about the puppy and his background so I could better understand where this puppy was coming from.

What I found out was the husband wanted the dog from the beginning. The wife was never in favor of getting a puppy because she did not want the responsibility of having to raise one. The husband decided he was going to get a puppy anyway. He purchased a seven-month-old puppy from a well-known breeder without his wife even knowing about it. (Read the section on puppies that come from breeders at this age at the beginning of this chapter.)

Ten seconds after I sat down, the dog lunged at the man's nine-year-old daughter. He could not reach her, so she was safe, but the dog was disappointed. Because he could not reach the child, he started biting at his owner's legs. I had the owner step on the leash to shorten it a bit more so the puppy could not bite his leg. He tried to bite the owner's leg again, so we shortened the leash once again, giving him just enough leash to move without being able to make contact with the man's leg.

The puppy, now not being able to lunge or bite, started barking profusely. Again I asked the father and little girl to completely ignore the barking. We talked about the show, the little girl's school, and many other things until finally the puppy decided to lie down. After a few minutes, I asked the owner to gently lower his hand and say in a soft voice, "Yes, good, quiet." In a spilt second, the dog was back up on all fours, so I asked him again to quickly ignore his puppy.

A minute or so went by, and the puppy finally laid down again, let out a big sigh, and laid his head on the floor. We waited about 30 seconds, and again I asked the owner to gently reach down and pet his puppy on the side softly while saying, "Yes, good, quiet." This time, the puppy did not even lift his head up; he just laid there quietly for the next few minutes.

After another five minutes, we were told it was time to go on the set. After I was introduced, the little girl was invited onto the set. She came on stage with her dog and told the audience how he jumps, bites, and knocks her down sometimes. When the host asked the little girl if she could demonstrate what that looked like, she stood up and the dog calmly walked with her and she gave him a pet.

The dog clearly was expected to jump, bark, bite, or do something obnoxious, but instead he was honestly being a good boy. In those few short minutes before the show, the puppy had an opportunity to learn a little about what was expected of him. The puppy was not trained in this amount of time, but he certainly understood that what he did before did not work. He knew if he wanted any attention, he would have to settle down to receive it. He learned a little self-control that day. I doubt if the behavior simply stopped, never to appear again, but at least the client knew how to settle him down in the future.

Another way we can work with hyperactive puppies is to put them in a crate and wait for them to settle down. Remember the client should use a leash, if necessary, to put the puppy in the crate or, if safe, pick the puppy up and place him in the crate. The client should not pull the puppy into the crate.

Have the client stay close to the crate so she can mark and reward the puppy when he is being quiet. Open the crate door slowly so that if he starts getting too excited again, the client can simply close the door and sit down near the crate and read a book. Tell the client to not talk to or look at the puppy until he is quiet. Once the puppy has completely settled down, mark and reward the quiet behavior and again slowly let him out of the crate. If he becomes hyperactive again, have the client close the crate door and repeat the above procedure until the crate door can be opened and the puppy can walk out calmly.

T-shirts and touches can help these puppies calm down beautifully. Do some touches first and then put the T-shirt on the puppy; in minutes, you will actually see him calm down. You can review the information on T-shirts and touches in the section on Alternative Methods in Chapter 3.

These training methods work for many puppies, but not for all. Some may need medication to help them get through this behavior. Review the client handout on Hyperactive Puppies in Appendix D for additional ways to address this behavior.

Once you meet with the client and puppy, decide which methods would work best for both of them. Those are the methods you should demonstrate in your PBS. Remember to give clients an opportunity to repeat what you demonstrate so you can be sure they will be practicing at home correctly.

Jumping

Jumping is a behavior that is often encouraged by clients when the puppy is young. As the puppy grows in size and weight, though, clients get angry at their puppies for jumping on them and their guests. This is an area where many young dogs get confused by clients, and it is a major reason that many large breeds are later surrendered to shelters.

I once had a conversation with a man who was in charge of all the animal shelters in the state of Michigan. He told me they euthanize about 400 dogs a week just because they cannot place large dogs that jump on people. Some of these dogs actually bit family members in the face accidentally while jumping on them.

This is one of those behaviors that can honestly mean life or death for many of our four-legged friends. It is a good idea to start clients off with this general rule: Never pay attention to a puppy that does not have all four feet on the floor. If we do this, we can stop the behavior before it gets out of hand and another dog has to lose his life or his home because of confusing signals and miscommunication from humans.

If a puppy is six months of age or older and is already a pretty consistent jumper, explain to the client how you will be addressing this behavior first. Let the client know you will need her help in working with this behavior. You should be in a private room when conducting your training sessions to address this behavior. Here are the steps you will take:

1. Explain to the client that when the puppy jumps up on either of you, both of you will walk quickly out of the room and close the door without saying a word to the puppy.

2. Ask the client to re-create a situation that would normally cause the puppy to jump up.

3. The second the puppy jumps up, both of you quickly walk out of the room and close the door, leaving the puppy isolated.

4. Wait 15 seconds and go back into the room and act like nothing bad happened. Greet the puppy in a happy voice. If the puppy jumps up, repeat the isolation exercise. You and the client may need to repeat this several times until the puppy can keep all four feet on the floor when you both walk back into the room. This may take five or ten attempts until the puppy figures out that when he jumps on people he is left all alone.

If the puppy catches on quickly, step out of the room and leave the client and patient alone in the room. Ask the client to again re-create the jumping behavior and to leave the room quickly without saying a word to the puppy if he jumps up. Once you feel the client understands how to work with correcting this behavior, you can end your session and schedule a second appointment. If time permits and you feel the client can handle more information, you can add the next step for training this behavior. No matter how smart the puppy is in your training session, the training is not complete because the correct behavior has yet to be established.

The next step for you or the client when working with the puppy's jumping behavior is to put the puppy on his leash. Stand on the leash, giving the puppy about three feet of leash from the clip on the collar to your foot on the floor, and elicit the jumping behavior again. The second the puppy tries to jump up, cross your arms over your chest, and ignore and look away from him. He will self-correct with the shortened leash. The second all four feet are back on the floor, mark and reward his four feet being on the floor with an ear scratch. The PuppySmarts Jumping training video demonstrates this nicely and may be a good tool to have the veterinarian prescribe when he or she joins your PBS at the end of the session.

When a new client does not have a leash on her puppy and the puppy tries to jump up, have the client cross her arms over her chest and look away from the puppy. This will protect her face and fingers. Once all four paws are on the floor, have the client mark the correct behavior and again give an ear scratch as a reward. If the puppy will not put all four feet on the floor, there are a few options to get the puppy to do this: (1) step into the puppy, (2) step back from the puppy, or (3) do a little leg wiggle to get the puppy off your leg. These movements can cause the puppy to gently lose his balance and cause him to want to put all four paws on the floor to regain balance.

I do not recommend giving a treat as a reward for this behavior until the client and patient are close to the finished behavior. You do not want to condition the puppy so that every time he jumps on someone he gets a treat. We want the conditioning to reinforce that when he jumps on people he is ignored. There is a fine line between ignoring an inappropriate behavior and rewarding it. For this behavior, until almost finished, the reward is the client's attention or an ear scratch. An ear scratch is still a reward, but not a high-value reward like a treat would be.

Grabbing paws, holding them, or squeezing them are not recommended ways to address this behavior, nor is kneeing the dog in the chest. These methods can encourage additional behavior challenges. For example, how do you think the puppy would then react to giving you his paws for nail trimmings? If you or the client are grabbing or squeezing his paws, do you really expect him to now trust you? Again, by using these extreme methods you may open the door to additional behavior problems, which neither you nor the client want to create.

Another issue with kneeing dogs in the chest is potential injury. Ask some of the veterinarians in your practice about the injuries they have seen from clients or trainers kneeing a dog in the chest. You will hear about dogs that came into the practice with broken collar bones and cracked ribs—two things you do not want to see happen to your patient.

Depending on the age of the puppy and how long he has been jumping on people, explain to the client what an extinction burst is so the client does not get discouraged in working with this behavior. If you are not clear on the definition, you can review the definition of extinction bursts in the section on Training Terminology in Chapter 3.

As always, end your training session on a positive note, even if that means asking the puppy for a *sit*. At least he will then have all four feet on the floor for you to mark and reward.

Leash Pulling

Puppies need lots of exercise, and going for a walk is one way a client can exercise the puppy. However, when the puppy is pulling on the leash, it can make it difficult for many clients to walk their dogs. As a result, the client will not be as willing to walk the puppy if she has to deal with the puppy constantly pulling.

Remember this basic rule: If you do not want your puppy to pull on a leash, then do not pull the puppy when on a leash. Instead, encourage the use of targeting, lures, and rewards to get the puppy to move forward, or away from things he should not be near. Refer to the client handout on Leash Pulling in Appendix D.

There are several reasons why leash pulling may be presenting itself. The first is my number one issue—extended leashes. This type of leash actually trains the puppy to pull on the leash to get more leash freedom. As the relationship unfolds and the puppy grows in size, strength, and weight, the pulling increases to extend the leash further. However, now it includes the client's arm and possibly the entire body being pulled by the puppy to extend the leash he has been taught to pull on for more freedom.

Another reason puppies learn to pull on their leashes is because they want to get from point A to point B at a faster pace than the client is walking. If the client allows the pulling to achieve the puppy's goal, she has just given her puppy a very good reason to pull on the leash. When the puppy begins to pull on the leash, the client should simply stop walking until the puppy stops pulling and turns his attention toward the client. Once the puppy stops pulling and starts paying attention to the client, the walk can proceed. Over a few days, the puppy will begin to realize whenever he pulls on his leash, all forward movement stops.

Training any dog with a six-foot leash gives the animal an opportunity to sniff the ground and walk a few feet away from the client, making the walk more enjoyable for both. Using a six-foot leash also gives the puppy a better opportunity to understand how far he can walk away without pulling on his leash and losing his ability to enjoy forward movement.

Leashes should be used as a safety precaution, not a crutch. The easiest way to train a puppy not to pull on his leash is by training the *target* cue. Once the puppy understands what the cue *target* means, leash pulling can cease. Here is how this can be used in working with puppies that pull on their leashes.

When you or the client in a PBS puts the leash on the puppy to go for a mini-walk, the first time the puppy gets almost to the end of his leash, give the cue "Target." The puppy will usually turn back toward the client to *target* the client's hand or a targeting stick, if he knows what the cue *target* means. For the puppy to *target* for the client, he will need to come back to her. Once the puppy *targets* the hand or stick, the client can mark and reward him with an ear scratch. Now that the puppy is closer to the client, the walk can continue.

Another way a client can work with leash pulling is to give the puppy treats when he is by the client's side. This teaches the puppy that when he stays close, he gets lots of treats. Over time, he becomes conditioned to walking close to the client without the need for treats.

The client can also tether the puppy to her waist while walking around the house. This accomplishes two things. First, it lets the client keep a closer eye on the puppy to prevent house-soiling accidents, and second, it teaches the puppy to walk within six feet of the client. Again, after time this becomes the conditioned place for the puppy to walk while on leash.

Exercise is very important to the young dog. If the puppy has been in a crate for six to eight hours, he needs some exercise and playtime prior to going for a walk. When the client lets the puppy out of the crate after several hours, the puppy will need to eliminate. Once that is done, playing fetch or tug-of-war for 10 to 15 minutes will help burn off a little excess energy before going for a walk.

When the puppy pulls on his leash all the time, the client will in many cases end up taking the puppy for fewer and perhaps even shorter walks. Then the puppy does not get the exercise he needs, and other behavior problems can occur that can strain the client-patient relationship. If all of the above-mentioned methods do not work for your client, there are training tools, such as halters, harnesses, or head collars that you can use. The question then becomes: Which tool would be best for this client and patient?

Head Collars

Different head collars offer different solutions for the patient and client, as mentioned previously. All of the products must be adjusted to fit the puppy properly and checked frequently, as puppies grow quickly. Many puppies accept head collars easily; others may need to be desensitized to them.

To desensitize the puppy to the head collar, you will need many tiny treats to distract the puppy. Most of these products come with directions on how to fit, adjust, and desensitize the puppy to them.

Once the head collar is fitted for the puppy, you can demonstrate how to use it. (*Note*: You want to train clients not to jerk the leash with head collars. They do not work like choke collars, and the puppy's neck could get hurt if the leash is snapped or jerked.) These collars almost work by themselves and need little effort from the client. All the client needs to do is stop walking and the head collar does the rest. You will find a few you can choose from in Table 4.1.

Table 4.1 Head Halters

Halti	**Halti**® is sold through Campbell Pet Products (www.campbellpet.com). The Halti fits looser on the dog than products such as the Gentle Leader® (see below) and does not have the clamp under the muzzle.
Snoot Loop	**Snoot Loop**®, made and sold by Snoot Loop (www.snootloop.com), is an excellent choice for some dogs. What makes this product unique is that it can be adjusted on the sides and back to accommodate tiny, short, and wide muzzles.
Gentle Leader	**Gentle Leader** is sold through Premier Pet Products (www.premier.com). It should be fit around the neck high and tight and has a clamp under the muzzle that must be adjusted a few inches under the muzzle so the puppy can breathe, drink, bark, and eat. Otherwise, it is similar to the Halti. Since puppies grow quickly, this product must be adjusted very frequently.

Halters/Harnesses

Since some clients and puppies resist head collars, halters or harnesses can be an excellent way to control leash pulling. Plain halters for dogs are sometimes made in a way that encourages the dog to pull. When shopping for halters, make sure the clip that will be attached to the leash is either on the chest or in front of the shoulders. Halters that attach to the leash behind the shoulders may cause some dogs to want to pull on the leash.

Some available halters and harnesses are shown in Table 4.2.

Table 4.2 **Halters or Harnesses**

The Sporn Training Halter	The **Sporn Training Halter™**, distributed by the Sporn Company (www.sporn.com), is very easy to put on and take off and is padded to be comfortable.
The Sporn Non-Pulling Mesh Harness	The **Sporn Non-Pulling Mesh Harness**, also distributed by the Sporn Company (www.sporn.com), is a continuation of their original training harness. It is also very easy to put on and take off and is padded to be comfortable.
The Sense-ation and Sense-ible Harnesses	The **Sense-ation™ and Sense-ible™** harnesses are distributed by Softouch Concepts (www.softouch-concepts.com). Sense-ation was the first harness of its kind to control leash pulling. Sense-ible is the later version of this harness, with the same design, but different material. The harness clips in the front of the chest so the dog will turn toward you when he starts to pull.
The Easy Walk Harness	The **Easy Walk™ Harness** from Premier Pet Products (www.premier.com) works similarly to the Sense-ation harness and is color-coded to make it easier for clients to remember how to put it on.
The Halti Harness	The **Halti® Harness** is sold through Coastal Pet Products (www.coastalpet.com) and made by Roger Mugford. This product requires two hands when you first use it, but once you master the use of this product, you can use it with just one hand.

Most major trade shows have many of these products for sale and you will probably find new ones coming on the market. The next time you attend a trade show, look for distributors that sell training products. Ask for information about these products and others they may carry. Your practice may already have information on some of the above-mentioned products, but perhaps not all of them, so check around and find the products you feel most comfortable adjusting and recommending to your clients. It is always a good idea to have many different tools in your toolbox of training products to choose from. Some puppies will work better with one product, whereas others will do better with another one. Likewise, some clients will work better with one product than another. Always choose your products based on the animal and client with whom you are working. Make sure at the end of your PBS that

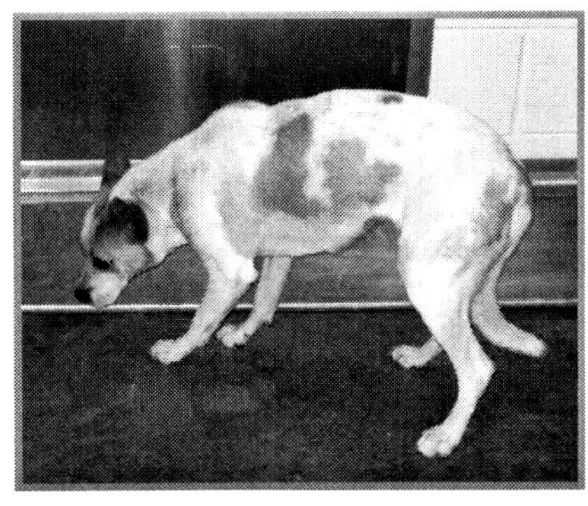

the doctor is the one to prescribe the leash-pulling product.

Shy, Timid, and Fearful Puppies

In many cases, puppies become shy, timid, and fearful as a result of their environments. Research done by Scott and Fuller[7] revealed that a dog's personality is 35 percent hereditary and 65 percent environmental. Based on these findings, there is a good chance that behavior is related to the puppy's environment. This behavior in many cases can be addressed by building the puppy's confidence. There are several ways you can do this.

Two very important things you will want clients to do with these puppies are to socialize them and also find out what stimulus the puppy is concerned with so desensitization can begin as well. It is extremely important that both you and the clients move slowly when working with these puppies. Do not force, pull, or demand anything of them unless you have to. Socialization and/or desensitization should always be done at the puppy's pace to build confidence.

Wearing a collar, walking on a leash, or allowing anyone to pet these puppies can be difficult for many of them. Again, the best thing we can do is to let these puppies experience anything new at their own pace. If the puppy pulls away from you or anyone else, you and the client must tell the person to please stop moving toward the puppy and/or stop what they are doing or trying to do. If the puppy freezes, stop what you are doing. The faster you stop, the easier it will be for the puppy to begin to trust you or anyone else.

One thing we have in our favor is that our dogs are very good at trusting us if we let them know quickly that we are listening to them by watching their body language. Stopping when the puppy shows concern lets him know you are paying attention to his feelings.

Clients need to begin socializing and/or desensitizing these puppies slowly. They should be carried by family members as little as possible! When puppies are constantly carried by their humans, the humans are sending a clear message to the puppy that he cannot handle anything on his own.

The puppy and client may enjoy this carrying around now, but by continuing in this direction, this wonderful puppy may end up with many additional behavior problems. For example, fearful, shy, and timid dogs may respond to life by shying away from unfamiliar people, sounds, or situations. Some may cower, bark, or snap at people without warning. Others may freeze in place. Some may try to fight with anyone or anything new or different. And still others may try to flee and run away from whatever concerns them.

Clients sometimes misunderstand their puppy's behavior and believe the puppy is trying to protect them, but this is rarely the case. More often, the puppy has learned that when he barks, snarls, growls, or snaps at someone, they will leave him alone.

This type of reinforcement can lead to aggressive behaviors if not addressed early in the puppy's life (before the age of 12 weeks).

One of the ways a client can work with such a puppy is to allow him to experience life with all four feet on the ground, even when larger dogs, strangers, or children are walking by. If the puppy seems concerned with the people or other animals, it is better for the client to cross the street than pick the puppy up. The client can start taking the puppy to more places and letting him just experience and watch the world from a safe distance.

After a week or two, the client can take the puppy closer to where people are walking. Have the client bring some of her puppy's favorite treats, and let strangers offer the puppy a treat. If the puppy backs away from strangers in the beginning, the client should stop the stranger from trying to give the puppy a treat or trying to pet him. Thank the person, and take the treat back so the person is not stuck holding a dog treat.

Never make these puppies take treats or receive pets from strangers. You do not want to make them more fearful; you are trying to build their confidence. If you force the puppy to do anything that he does not want to do, you defeat the purpose of the whole exercise.

Share with clients they should pay less attention to the fearful behavior, and instead mark and reward his brave behavior.

While introducing the puppy to others, have the client pay close attention to any behavioral patterns they may see. For example, is the puppy more concerned with people wearing glasses, wearing hats, or with men, women, or children? This will help the client understand her puppy's desensitization needs. When desensitizing her puppy to any stimuli with which the puppy shows concern, it is very important not to flood him. (If you are not sure what flooding means, read the definition in the section on Training Terminology in Chapter 3.) Flooding can make matters worse. Small steps when working with these puppies are always best.

Another technique that can be very helpful in working with timid, fearful, and shy puppies is target training. Refer to the section on Target Training in Cues/Commands. This can be used to shape many behaviors and help overcome fearful/shy behaviors. It gives the puppy something to focus on when a challenge is presented. Once the puppy is comfortable with targeting, he will focus on the target more than the new experience. It is something he can rely on to feel more comfortable as he faces new obstacles and challenges in life.

Have the client introduce the puppy to new things every few days while at home. She can use a broom on the floor for the puppy to walk over, a pool noodle, stairs to go up and down, and so forth. Tell the client to use her imagination. The vacuum cleaner should be saved until the puppy has a bit more confidence, as many dogs are concerned about noisy things. It is definitely not the first thing with which you want to try to get the puppy comfortable. When the vacuum is introduced, it should be set in the off position.

Once the puppy is comfortable with a few things inside the house, the client can start to introduce him to the world in which he lives. Tell your client to just take it slowly. When on outings, it is a good idea for the client to bring lots of tiny treats so others can offer treats to the puppy. If the client takes her time in working with the puppy, the puppy's confidence level will begin to increase.

The client can take the puppy to a shopping mall and sit outside on one of the benches as far away from the door as possible. This way the puppy is not close to the action of the entrance, but can observe what is happening. The assignment is to do this a few times a week, each time moving closer to the entrance. However, the same rule regarding strangers holds true: If the puppy is willing to accept a treat or a pet from strangers and walks up to them, that is fine. If the puppy backs away, the client should thank the strangers for trying and politely ask them to walk away.

Another training tool you can use with these puppies is touches and T-shirts. Information on both of these methods can be found in the section Alternative Methods in Chapter 3. Using touches can help animals move past habitual patterns to relieve stress, allowing the puppies to think differently instead of simply reacting. T-shirts give puppies a better understanding of their own bodies and can comfort a shy, timid, or fearful animal.

Being afraid in life does not feel very good for anyone, animal or human. Keeping that in mind should help you and your clients to be patient and understanding with these puppies. Desensitization, counter-conditioning, and building confidence are all very useful in helping these puppies overcome their fears. Review the sections Training Principles for Puppies and Kittens as well as Desensitizing and Counter-Conditioning to Overcome Negative Reactions in Chapter 3. Share the stories in those sections with your clients to help them better understand these behaviors and have a little more patience with their shy and timid puppies.

So when working with shy, timid, and fearful puppies, take your time. You are not in an exam room under pressure to do anything quickly. You are in the PBS to help a client and patient be successful together.

In your PBSs, you can cover a few ways clients can begin to socialize and/or desensitize their puppies. In addition, you may want to demonstrate what target training looks like to give the puppy something to focus on. You can also offer a few suggestions on how and where they can begin this type of training. The information can be found in the sections of Chapter 3 referred to above as well as the client handouts on Shy, Timid, and Fearful Puppies.

Endnotes

1. John Paul Scott and John L. Fuller, Genetics and the Social Behaviors of the Dog. (Chicago: University of Chicago Press, 1998).

2. D. G. Freedman, J. A. King, and O. Elliot, "Critical Period in the Social Development of Dogs," *Science*, 133 (1961): 1016-1017.

3. Scott and Fuller, 1998.

4. http://www.purdue.edu/UNS/html4ever/1998/9802.Luescher.puppies.html, Accessed November 14, 2008.

5. Ibid.

6. These and many more bite statistics are available at www.cdc.gov and www.dogbitelaw.com.

7. Scott and Fuller, 1998.

CHAPTER 5:
KITTEN BEHAVIOR SESSIONS

Understanding the Kitten's History

Appendix B contains a Kitten Behavior Session Checklist to guide you with your Kitten Behavior Sessions. The general section of this checklist presents important questions about the kitten's background and her new family structure. The kitten's history will give you a better understanding of the kitten.

The first set of questions relates to where the kitten came from. This will let you know whether the kitten came from a reputable breeder, a pet store, a shelter, a backyard breeder, or if the family found the kitten or the kitten found them. Each scenario may present unexpected challenges.

Pet Stores

Relatively few kittens are purchased from pet stores, but there are some. "Kitten mills" probably do not exist because a kitten that is not handled at all when young would be feral. Feral kittens are not easily integrated into family life. In most cases, the pet store probably deals with a local cat breeder. As with all breeders, the amount of time spent socializing the kittens to people, litter boxes, touch, and toys is important. Find out if the client knows how long the kitten was at the pet store. A few days will probably not cause much damage. However, a few weeks could cause many challenges. The kitten may be fearful if she has been handled by many strangers. She may also be timid and/or shy, wanting to stay away from people. She may have litter box problems, depending on the amount of cleanliness and privacy she was given at the pet store. All of these situations can affect her view of the world and how she handles it. Time, patience, and consistency from the client and you will be very important to these kittens.

Kittens from Professional Breeders

Breeders who are aware and conscientious about their kittens' socialization needs usually have happy, normal, playful, and curious kittens. Any behavior challenges the client is experiencing with a kitten from a professional breeder are probably normal behaviors a kitten goes through, and the client just needs a little help addressing them. Play and socialization can do wonders for behavior challenges with kittens.

In the search for the perfect "show cat" that will be a consistent winner in the ring, some breeders will take shortcuts that can emotionally harm the kitten. Pushing too hard or too fast can cause a variety of behavior issues. If a cat develops behaviors that make her unsuitable for showing, the breeders will try to sell the pet. Depending on the behaviors, the cat may have "special needs," which can make her more challenging as a family pet.

Shelters

Consider the many reasons a kitten is at a shelter in the first place. The kitten may have been dropped off because the mother died during the birth and the owner did not want the responsibility of raising kittens. The litter and mother may have been dropped off because the owner did not want the litter. The kitten might have been found outside. The kitten may have had a home and has already lost it. In many cases, the kitten was not given the chance to be successful. For all you know, this kitten could have had two or three homes before your client's.

Many people believe cats do not have to be trained and that cats like to be left alone. These statements are not true. Cats, like dogs, do need training, and they enjoy and need social time with their family. Kittens also need and deserve to have their play drives met.

Many shelters lack the space a kitten needs for privacy and exercise. This lack can be the cause of the behavior problems your client may be facing with a newly adopted kitten. If the client is patient with the kitten, in time many behavior challenges will stop if the kitten's basic instinctual needs are met.

Backyard Breeders

Backyard breeders range from excellent to poor. The backyard breeder may have raised the kittens in a stimulus-poor environment or stimulus-rich environment. For kittens from backyard breeders, you could add a few questions to your questionnaire: Were there children in the home where the kittens were born? Did the children have access to the kittens? Was the access supervised? To the best of your knowledge, do you believe the kitten may have been abused or mishandled by unsupervised children or uncaring adults? How much were the kittens handled and socialized?

Some backyard breeders are excellent caregivers that handle and socialize their kittens by the book. When kittens from good backyard breeders are in a Kitten Behavior Session, the client may need just a little help to get the kitten on track.

Understanding Kitten Development Stages[1]

Kitten desensitization and socialization start shortly after a kitten is born. Kittens seem to develop much faster than puppies. Kittens go through many of the same stages as puppies, although the exact timeline is not as clear as it is with puppies. The following timelines are approximate, not exact. Keep in mind that many people will not get a kitten until she is eight to twelve weeks old, when she is weaned from her mother.

Neonatal Period: Birth to One Week

When kittens are just a few days old, desensitization can and should begin. At this stage, the kittens should be picked up and held gently one at a time as one would a human baby. The kitten should be held in one's arms while her stomach is gently rubbed. The kitten's paws, face, and ears should be lightly touched for just a few minutes a day to stimulate the nervous system and desensitize the kitten to being held and touched. Needless to say, at this stage the kitten is extremely delicate and should be treated much as a newborn baby would be treated.

Transitional Period: Two to Three Weeks

During this time, a kitten's eyes open and she begins to explore her new world.

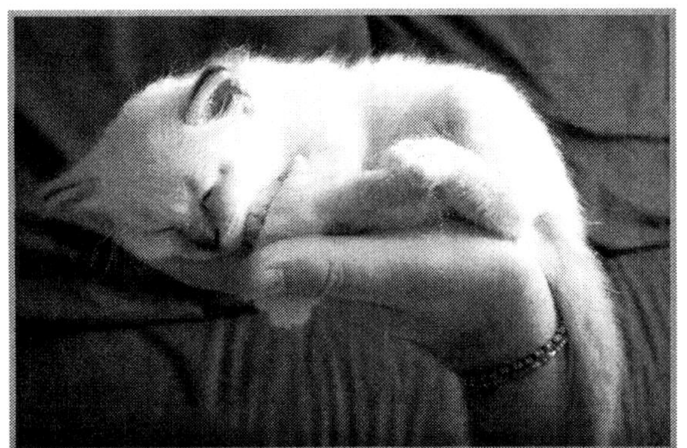

Motor and sensory skills are developing at this time. Continue picking the kitten up daily, rubbing and touching her all over. Extend the amount of time each kitten is held. Instead of just a few minutes a day, do these exercises many times a day and for longer periods of time. Although the kitten has increased in strength a little, clients should still be aware that the kitten is still very delicate.

Until the kitten is three weeks old, children should not be allowed to play with the kittens. If a child wants to pet or touch one of the kittens, the parent should hold the kitten and supervise the interaction.

Socialization Period: Three to Nine Weeks

When kittens are handled gently by people as early as a few days old, they develop faster physically and their social interactions with humans are improved. The sensitive period for socializing kittens to humans runs from two to seven weeks of age. During this time, the kittens should be introduced to and held by many people. Children can now hold the kitten, but they should be supervised when doing so. This is also a great time to introduce the kittens to stimulating toys and a few scratching posts. Other safe animals should be introduced to the kittens during this

period as well. The daily handling and touching of each kitten should continue.

Desensitizing the kitten to a cat carrier during the socialization period will make visits to the veterinary practice and travel easier. Some clients will enjoy leash training their kittens and taking them out to as many places as puppies go. Carrier training should continue from around the third week of age until adulthood.

Litter boxes should also be introduced during this socialization period. Finer litter material just an inch or so deep is a good start. Boxes should be cleaned often throughout the day. The box should be washed out at least a few times a week, or more frequently if necessary. Cats are very sensitive to odors that we cannot even detect, so clients should not use harsh chemicals to keep the litter box clean.

Kittens of any age should not be teased or roughhoused with human hands, nor should kittens be allowed to scratch or bite at any age. If the kitten tries to bite or scratch, the client can make a loud noise, such as a clap or a snap, or loudly use a word, such as "ouch," to interrupt the behavior. When a scratch or bite occurs, all play and attention stops. Clients should only use cat toys, not hands, when playing with their kitten. This will prevent future accidents from occurring.

Kittens should not go to their new homes until they are at least eight weeks old. The following developmental stage is when many of your clients will be beginning their relationship with their new kittens.

Juvenile Period: Ten Weeks to Twelve Months

This period is when a kitten's independence begins. If a kitten has not had socialization and desensitization opportunities to a large degree by this point, the kitten

can be feral. Adjusting a feral kitten to a home can be a long process, but it is not an impossible endeavor. With time and patience, these kittens can be tamed. However, this must be done slowly and at the kitten's pace. You will always want to build on the animal's successes.

Knowing how kittens were raised during the early stages of their lives, as well as the socialization and desensitizing opportunities provided during these stages, will give you a good indication of what a kitten will be comfortable with as an adult cat. It is important to recognize that many kittens do not grow up in nurturing environments during their early stages of development. This can cause a kitten to be more withdrawn and antisocial when being introduced to her new family. It is very important to remind your clients that a kitten should never be forced to do anything. This makes it difficult to develop a close, trusting bond between your client and the new kitten.

Kitten Training Basics

Training a kitten is a long process and can become very frustrating for your client. Many clients may believe a kitten does not have to be trained, but this is not true. Although their training needs are different from those of a puppy, kittens still need training. Kitten training includes litter box training, socialization with people, and other animals, and desensitization to stimuli that concerns them. In addition, kittens need to learn that biting or scratching humans is an unacceptable behavior.

Training a kitten can take much longer than training a puppy, but both animals learn the same way. Most cats seem to respond better to clicker training than to verbal cues. That is not to say verbal cues do not work, because they do. For more information on clicker training, refer to the section Training Methods in Chapter 3. Kittens move on to the next interesting thing so rapidly that many opportunities for rewarding appropriate behavior can be missed. Marking and rewarding appropriate behaviors quickly is especially important for a kitten.

Things to remember when training a kitten include the following:

- Always work at the kitten's pace.
- End the training session before the kitten does or she will get bored and lose interest.
- End all training sessions on a positive note; a long, soft pet or a small treat can be wonderful ways to end a Kitten Behavior Session.

Kitten Socialization

Review the section in Chapter 3 titled Socialize for a Confident, Well-Adjusted Pet as well as the definition of socialization under the section in Chapter 3 titled Training Terminology before reading this section. The following discussion applies to kittens only.

The window of opportunity for a kitten's socialization is much larger than a puppy's and occurs between the ages of two weeks and twelve months. A kitten must become desensitized to the cat carrier as well as all the other stimuli she is likely to encounter with her new family.

Environmental stimulation should be a part of every cat's life. Without this stimulation, behavior problems can—and in many cases will—occur. Cats are curious creatures and will investigate any changes to their surroundings. Just like dogs, cats need their toys rotated to keep them interested. They need all the aspects of environmental enrichment: olfactory (smell), gustatory (taste), hunting, hiding, predatory, and social. Some ideas on how clients can offer their kittens the stimulation they need are listed below:

- *Olfactory*. Make different scents available. Cats love to smell things. Cat grass is available at most pet stores and offers the kitten something different to smell, as do other safe houseplants. Make sure the client does not have poisonous plants in the home. For a list of poisonous plants, go to http://www.cfainc.org/articles/plants.html.
- *Gustatory*. Cats enjoy different flavors and different textures. Eating the same thing all the time can become very boring to a cat. Offering a kitten some canned food once a week can be a nice change. Varying cat treats will help the kitten look forward to training sessions.
- *Hiding*. Paper bags, kitten play tunnels, and cardboard boxes are some favorite hiding places for kittens. Other favorites include cabinets, behind couches, or under beds.
- *Predatory/Hunting*. Because cats are predatory animals, they love toys they can chase, pounce on, or grab. Provide toys for the kitten to try and capture,

such as a feather, a tasseled wand, a laser pointer, or moving objects that she can chase. Always make sure the client completes the predatory sequence when they are done playing by allowing the kitten to capture the toy. In the case of a laser light, point the light on a toy she can "kill." The sequence must be completed or she may become frustrated and other inappropriate behaviors can manifest.

- *Social.* Interactive safe play with family members provides socialization for the kitten as it does for other animals in the household. A good kitten is a kitten that has many opportunities to get her needs met: physically, mentally, and emotionally.

Collars, Harnesses, and Leashes

Some clients may want to teach their kittens to walk on a leash. Use a cat harness instead of a simple collar. The kitten first needs to be desensitized to the cat harness. It is normal for kittens to object when they begin wearing a harness. It is something new they are not used to. If a kitten is reacting badly to her collar or a harness, you and the client will need to gradually desensitize her to the feel.

Let the kitten smell the harness first. Then place the harness on the kitten's back to let her get used to the feel of it. While you are offering her treats, ask the client to put the harness on the kitten. If the kitten tries to escape, hold her for just two or three seconds and let her go. Wait a few minutes and repeat the above exercise until the client can put the harness on the kitten without her trying to escape. Once the harness is on, give the kitten some time to get used to the feel of this strange object now attached to her. After three to five minutes, depending on the kitten's reaction, take the harness off and give her a jackpot of treats. Take the harness off sooner if she reacts badly, and start again in a few minutes.

Even more so than puppies, kittens tend to dislike leashes, but they can become desensitized to walking with one. Since cats seem to respond to clicker training very nicely, first condition the kitten to the clicker. Once the kitten has been conditioned to the clicker, put the cat on a leash, and use a dangle toy to entice the kitten to take a step forward toward the toy. The second she takes her first forward step, click and reward her first step on the leash. Repeat this several times over many days, and soon she will take two, three, or four steps while on her leash. Some kittens accept the leash quickly, but the majority of kittens do not. Cats can be trained to walk on a leash, but it is a slow process for many cats.

Exercise and Play

Exercise and play are important for all kittens and adult cats. Exercise for a kitten is more than her walking through the house with the client. Play is more than giving her a toy and leaving her to play by herself.

Kittens have different play drives. These instinctive drives must be nurtured. If the kitten is left alone to play by herself all the time, inappropriate behaviors will manifest themselves for lack of adequate stimulation. When clients take the time to exercise and play with their kittens, addressing their scratching, territorial, predatory, and locomotive needs, they will have a happy, well-adjusted kitten who will become a wonderful family member.

Exercise develops kittens physically, and play helps develop the kitten socially and mentally. Exercise consists of activities that allow the kitten to draw on her natural instincts, and build strength, agility, and coordination. Play is the interaction between the kitten and other family members or other animals. This is how kittens build social skills and learn how to interact properly with others. Many activities incorporate both exercise and play.

Kittens need to learn how to play safely with people and what is and is not acceptable behavior. Scratching, biting, and pouncing on humans are not acceptable ways for kittens to play with people.

For additional information, review the section on Playing Safely with a Kitten at the end of this chapter as well as the client handout Playing Safely with Your Kitten in Appendix E.

Kitten Behavior Challenges

Before offering corrections for any behavior, take note of the kitten's environment first. Does she have the toys she needs to entertain herself? Are her play drives being met appropriately? Does she have social time with her family? Is she left alone too frequently or too long? If the kitten is not having her needs met and she is bored, would it be appropriate to reprimand her? The answer is no. The kitten should not be reprimanded for simply being a kitten. If the client is not helping her meet her needs, the only one she can rely on to do so is herself. She will have to create her own stimulating environment, which may include creating her own games. Clients need to understand their kittens' needs and address them before ever giving a reprimand.

If a correction is in order, I have been more successful with simply shaking a plastic bottle with coins in it and saying a word such as "off." The second the kitten gets startled, she will usually stop what she is doing and escape, which stops the behavior. Then I would talk to her with a soft voice to let her know it is okay and that she was a good girl for leaving the item alone or getting off the countertops or whatever she was on.

Counter and Table Surfing

Surfing refers to kittens jumping on and/or playing on counters or tables. Most kittens will do this from time to time for a few different reasons.

It is natural for kittens and adult cats to explore their home, including countertops. Kittens are very curious animals; when there are objects on tables or countertops that look interesting to them, they will want to check them out. These objects can include photo frames, candles, coasters, and all kinds of things. For a kitten, all of these objects are possible play toys. If this is the reason the kitten is jumping on the table or counter, have the client kitten-proof the house.

Sometimes a new toy to play with may distract her temporarily, but kittens and cats enjoy high places to keep an eye on what is happening around them. If the client provides a cat tree that is higher than the countertops with a little catnip rubbed on it (to make her new tree more interesting), she will prefer the cat tree to the countertops. When a kitten is getting into a lot of mischief, she is probably bored and needs more stimulation in her environment.

Another reason a kitten may jump on a table or a countertop is because it stimulates her locomotive play drive. Cats like high places where they can easily watch what is going on. Ask the client specific questions on when the kitten is jumping up and what the kitten is attracted to that is on the countertop or table. When does the client notice this happening? How often? Is it one particular table or counter? Is there food on that table or counter? Are there plants, real or fake, on the countertop or table? These are all questions that may help you to determine the reason the cat is leaping up on these surfaces.

Remember, kittens love to explore, so to intervene on counter surfing the client needs a tall cat tree or trees. Have the client place these trees where the kitten will have a better view of what she wants to see from above, so the kitten no longer wants to view the world from the countertops. If the client purchases a three-foot high cat tree, the kitten will still prefer the countertop. If the client puts the cat tree in a corner where she cannot see outside, the kitten will probably still prefer the countertop.

Cat trees or climbing posts should be taller than the countertop and, if possible, taller than anything else in the room. Placed by a window, the tree will become a haven of enjoyment for the kitten. This will give her a vantage point to view the outside world while keeping track of what is happening in her territory. Have clients place the cat trees near windows and/or in a "busy" area of the house so the kitten can be entertained by watching the activity inside or outside.

Kittens who jump up are exploring their environment, which is normal. If a correction is necessary, simply shaking a plastic bottle with coins in it and saying a word such as "off" will do. For recommendations on how to diminish this behavior, refer to the client handout on Counter and Table Surfing in Appendix E.

Knocking Things Over

All kittens will accidentally knock things over from time to time. However, many kittens will stretch out their paws and purposely knock things down. It may seem like a strange thing to do; however, this again is due to curiosity and the kitten's play drive. Consider a stack of clothes, for example. The kitten is probably wondering what this thing is and whether it is living or something to be played with. The cat will knock over the pile of clothes and then most likely become a little confused. She may run in circles around the clothes, and then pounce on them. Again, this is her predatory play drive at work. Kittens need appropriate toys to stimulate their senses. When they do not get their needs met, behavior challenges may occur. Kittens, just like puppies, need to interact with their families often. Many kittens are very social, and when left alone for long periods of time, they get bored and look for something to do.

A few suggestions to give to your client to deal with this behavior include the following:

- *Provide stimulating toys* that the kitten can play with. Interactive toys that reward her with treats when playing with them will stimulate her mentally and physically. These are safe toys she can play with by herself. It is important to rotate her toys so she does not become bored with any of them.

- *Deter Bad Behavior.* When things have been knocked over, the kitten should hear a sudden sound or be sprayed with a water bottle as a deterrent. The water bottle is not a favorite training tool of mine, but in many cases it does work. If the client does spray the kitten with a water bottle, be careful. The kitten needs to associate a spray of water (something they do not like) with the act of knocking something over, and not with the client.

- *Cat-proof the house.* Put items the kitten has previously shown an interest in away so she cannot get at them again. Have the client purchase a sticky product to use under items she wants to leave out. The client may also consider getting heavier knickknacks to leave out so they are not easily knocked over by the kitten.

- *Add a tall climbing tree* in a location where the kitten can look outside and still keep an eye on her territory. Rub catnip on the climbing tree to make it smell wonderful so the kitten will want to check it out.

- *Recommend more interactive playtime* to meet the kitten's instinctive needs and play drives.

Litter Box Troubleshooting

Kittens do not usually have to be trained to use a litter box once they know where it is and, of course, if it is in a convenient yet somewhat private location. However, some kittens will not use a litter box. If the kitten is eliminating inappropriately, there is probably a preference problem. In these cases, some of the tips below may help to determine the kitten's preferences. Wait a week or two between each change to allow the kitten to make a choice and show her preference.

Have the client put down two litter boxes or one per kitten plus one extra. Tell the client to clean the litter boxes at least once a day. It may be that the litter box is not clean enough for the kitten. With two litter boxes, the kitten has an alternative.

Kittens and cats like a clean litter box and clean litter. Even clump litter that can be easily scooped out of the box still needs to be changed at least once a month. Litter boxes must be cleaned out at least once a day. The litter box itself needs to be washed out monthly. If urine or feces gets on the sides of the box, the box should be cleaned immediately. Leaving a litter box dirty leads to elimination problems.

The size of the litter box matters to a kitten or adult cat. Each litter box should be large enough for the kitten to feel comfortable. A good rule is to use a litter box three times the length of the kitten from nose to tail. Young kittens and older cats need low sides to get in or out easily and comfortably. Many commercial litter boxes are too small for many cats. Sometimes, using a larger container can make all the difference because the kitten does not feel confined when using her litter box. Under-bed storage boxes work great as litter boxes and offer the kitten the room she needs to feel comfortable.

The kitten may not like the location of the litter box. The kitten will not want to feel trapped, so the area containing the litter box should have a few escape routes for the cat if startled. At the same time, kittens want some privacy, so clients should not put the litter box in a high traffic area of the house.

Cats normally do not like to share litter boxes. If the home has four cats, the client should have five litter boxes, or one for each cat plus a "spare."

Since many kittens naturally look for soft matter, such as sand, to eliminate in, they seem to prefer softer litters that clump. Other kittens may like smooth surfaces to eliminate on. Have the client offer two litter boxes, one with litter and one without. Have the client leave the boxes down for two weeks so she can determine the kitten's preference.

The amount of litter in the litter box is also important for the kitten. Some kittens may want just enough litter to cover the bottom of the box. Others may want two to three inches of litter in the box. Some may even prefer four to six inches of litter in their boxes. The client can use varying amounts of litter in the two litter boxes once it is determined that the kitten does prefer litter instead of smooth surfaces. The client should change the amount of litter in the box gradually and leave the box at each level for at least two weeks before adding more litter to it.

When the client is first introducing the kitten into the home, the litter box will be in the same area as the kitten. As the kitten has more freedom to roam, the client may wish to relocate the litter box. The client can place additional boxes in locations they would like the kitten to use, or slowly, a little each day, the client can move the litter box to a new location in the home. When adding or moving the litter box, it is important for the client to always keep in mind location, traffic, escape routes, and noise.

As with puppies, rewarding appropriate behaviors is very important, especially if the client is having house soiling problems. The client should pay close attention to the kitten so she can mark and reward when the kitten eliminates in the proper area when litter box training becomes a problem. Marks and rewards should be given when the kitten is *done* doing her business, not *during* the elimination process.

As an example, I once had a client who was having trouble with her cat eliminating outside the box. The client tried a different litter, a larger box, less, and finally more litter in the box. This cat seemed to like the deeper litter and used her litter box for the first time in months. The client marked and rewarded her cat with freeze-dried shrimp. Her cat responded quickly to the treats and would actually take long strides across the floor toward her litter box to get the client's attention on the way to the litter box. I believe the cat was informing her to get the treats ready. The treats, along with more litter in her litter box, seemed to do the trick.

Another reason a kitten or cat may not want to use the litter box is because of a change in the household or family. Cats at any age do not like changes. I have seen cats stop using their litter boxes simply because the client moved the furniture around in the home. Because cats are not generally fond of change, it can sometimes be very difficult to figure out why the cat has decided not to use the litter box. Only through the use of probing questions can you find the reason why a cat/kitten has stopped using the litter box.

For example, one client of mine went on the Atkins Diet and stopped eating cereal in the morning—and her cat suddenly stopped using her litter box. The client used to eat her cereal in the morning and give her cat the remaining milk in the bottom of the bowl. The cat was showing that she did not like the idea that her morning milk was taken away. I suggested that the client give her cat just a half ounce of milk in the morning to see if that would make a difference, and it did. Her cat resumed using her litter box later that day. Some cats are lactose intolerant, and milk—especially cow's milk—is very difficult for them to digest. Obviously, this particular cat wasn't lactose intolerant!

Playing Safely with a Kitten

Kittens are very curious animals. They can become very vivacious, especially when investigating new objects. For this reason, a kitten should never be teased with hands

or feet. We do not want the kitten to learn that hands and feet are play toys. After all, it is not the human who will lose his home. A kitten does not discriminate between hands and feet and a toy mouse. Only appropriate cat toys should be used while playing with kittens or cats.

Cats have two primary modes of play: predatory and locomotive.

- *Predatory.* The play behaviors associated with this mode are pouncing, grabbing, chasing, and throwing things in the air.
- *Locomotive.* The play behaviors associated with this mode are running, climbing, leaping, and finding places they can go into and come out of quickly, like a paper bag or a box.

Kittens should be introduced to interaction within the household. One could regard this as a third type of play, known as social play.

- *Social Play.* This play is with other animals or people in the house. This is a great time to use interactive toys with the kitten so she learns to enjoy playtime with people.

Kittens also need toys they can play with by themselves. Among the best toys for kittens to play with unsupervised are toys that the client can put treats in for the kitten. These toys dispense pieces of food, offering the kitten mental and physical exercise.

Again, when the kitten participates in social play, it is important that hands are not used as toys. Toys are the safest way to play with kittens and cats.

There is no such thing as an accident. If the kitten scratches or bites the client, all play and attention should stop and the client should walk away from the kitten.

Pouncing on Legs

Some kittens enjoy jumping on legs when people walk by them. This can be part of their predatory play drive or because the kitten is trying to get the client's attention or is simply bored. If this happens, the client should make a loud sound such as "ouch" or clap her hands sharply, interrupting the kitten's behavior, and then walk away. The client may need to leave the room, especially if the kitten keeps on doing this. The kitten will eventually learn that legs are simply not fun to play with, and pouncing on legs does not get the attention they may sometimes desire. It is important that the client not play with or pick up the kitten when it jumps on her leg. If she does, she is rewarding a behavior she wants to stop, which will confuse the kitten.

The most common reason kittens pounce on legs or feet is because they are bored and their needs are not being met. Clients can enrich the kitten's environment by providing appropriate toys for the kitten to play with by herself. Spending time playing with the kitten and giving her an opportunity to use her predatory play drive will in many cases eliminate this behavior. When playing predatory games with a kitten or adult cat, it is important to let the pet catch the object and "kill" it. This gives the animal the opportunity to complete the predatory sequence.

Here are some general suggestions when working with this behavior:

- Make a loud noise, such as a clap or snap, to startle the kitten to interrupt the behavior.
- Mark and reward the kitten when the client can walk past her without getting her legs pounced on.
- Reward good behavior with a simple long, soft pet, a treat, or circular touches. Review Touch Work in the section Alternative Methods in Chapter 3 for more information on touches.

- Offer acceptable toys the kitten can walk in, hide in, climb, chase, and pounce on. When these items are not available, the kitten will be left on her own to make up predatory games, which can include legs.
- Make sure the kitten receives interactive playtime with her family members.

Scratching and Biting

Scratching and biting are normal behaviors for kittens when playing with other cats. The kitten must learn that human skin is more sensitive than hers, and scratching or biting us is unacceptable. All play should stop when the kitten scratches or bites. There are no "accidents." Kittens and cats know when their claws are in or out. They also know when their mouths are open or closed. So the first thing you want to help clients understand is the behavior is not an accident and needs to be addressed. When the client simply stops paying attention to the kitten and walks away, in time the kitten will learn she does not receive attention when she offers these behaviors, and any playtime that was occurring stops.

If scratching and biting occur seemingly out of nowhere, make sure to ask the client some questions that may help get to the root of the problem. Kittens are extremely sensitive animals, and small changes can disrupt them greatly. Make sure you ask specific questions on when the behavior started, who the behavior is directed toward, how many scratching posts the kitten has, and whether your client thinks it is play, affection, or aggression.

Kittens need a few scratching posts to scratch on. One in every room would be perfect. You will also want to know what changes have happened in the household. Ask the client what she thinks could be the cause. Does the kitten receive interactive playtime with her new family? How long and how often? What is her favorite toy? What is her favorite game? What is her favorite treat? Does she have toys that she can play with by herself that offer mental stimulation? Kittens need toys they can jump on, go inside of, bat with their paws, and climb on. Does the kitten have toys that will enrich her environment and satisfy her curiosity?

Kittens and cats need scratching posts throughout their lives. Ask the client how many scratching posts she has and where they are located. Does the kitten use the scratching posts? If not, perhaps the client needs to put a little catnip on the posts and reintroduce the kitten to the scratching posts or pads.

Kittens are often misunderstood. Many clients do not realize that kittens require training and socialization. They do not know about a kitten's play drives and sensory needs. If these needs are not met, behavior problems often will occur.

Endnote

1. G. Landsberg, W. Hunthausen, and L. Ackerman, *Handbook of Behavior Problems of the Dog and Cat* (Philadelphia: Elsevier, 2003).

Bibliography

American Veterinarian Medical Association, 2003 Survey.

Catanzaro, Tom. *Promoting the Human-Animal Bond in Veterinary Practice.* Ames, IA: Iowa State University Press, 2001.

Freedman, D. G., J. A. King, and O. Elliot. "Critical Period in the Social Development of Dogs." *Science.* 133, 1961: 1016–1017.

Landsberg, Gary, Wayne Hunthausen, and Lowell Ackerman. *Handbook of Behavior Problems of the Dog and Cat.* Philadelphia: Elsevier, 2003.

Overall, Karen. "Are you Encountering Problematic Behavior?" *DVM Newsmagazine.* February 1, 2005.

Overall, Karen. "Evaluation and Management of Behavioral Conditions." In: *Clinical Neurology in Small Animals— Localization, Diagnosis and Treatment,* edited by K.G Braund. Ithaca, NY: International Veterinary Information Service, 2001. www.ivis.org, Accessed June 2005. edited by K.G Braund. Ithaca, NY: International Veterinary Information Service, 2001. www.ivis.org, Accessed June 2005.

Overall, Karen. "Resolve Common Behavior Problems by Offering Training Advice to Clients." *DVM Newsmagazine.* June 1, 2002.

Salman, Mo D., Jennifer Hutchison, Rebecca Ruch-Gallie, Lori Kogan, John C. New, Jr., Phillip H. Kass, and Janet M. Scarlett. "Behavioral Reasons for Relinquishment of Dogs and Cats to 12 Shelters." *Journal of Applied Animal Welfare Science.* 3(2), 2000: 93–106.

Scott, John Paul, and John L. Fuller. *Genetics and the Social Behaviors of the Dog.* Chicago: University of Chicago Press, 1998.

Shore, E. R., and K. Girrens. "Characteristics of Animals Entering an Animal Control or Humane Society Shelter in a Midwestern City." *Journal of Applied Animal Welfare Science.* 4, 2001: 105–116.

Sousa, David A. *How the Brain Learns,* 2nd edition. Thousand Oaks, CA: Corwin Press, 2000.

Tremayne, Jessica. "AAFP Pens Behavior Guides for DVMs, Staff, Clients." *DVM Newsmagazine.* April 1, 2005.

Weiss, Harold B., Deborah I. Friedman, and Jeffrey H. Coben. "Incidence of Dog Bite Injuries Treated in Emergency Departments." *Journal of the American Medical Association.* 279, 1998: 51–53.

www.americanhumane.org/site/PageServer?pagename=nr_fact_sheets_animal_euthanasia. Data on file. American Humane Society 1997-2007. Accessed November 2007.

www.cdc.gov. Centers for Disease Control and Prevention. Accessed June 2005.

www.dogbitelaw.com. Accessed June 2005.

www.dogbitelaw.com/PAGES/Whipple.html. Accessed November 14, 2008.

www.flvetbehavior.com/Web/FAQ-Behavior_Professionals.html. Accessed March 2008.

www.iii.org. Insurance Information Institute. "Dog Bite Liability." Accessed January 2005.

www.phac-aspc.gc.ca/injury-bles/chirpp/injrep-rapbles/dogbit-eng.php. Canadian Hospital Injury Reporting and Prevention Program. "Injuries Associated with Dog Bites and Dog Attacks." Accessed June 2005.

www.purdue.edu/UNS/html4ever/1998/9802.Luescher.puppies.html. Accessed November 2008.

Suggested Reading

General

Landsberg, Gary, Wayne Hunthausen, and Lowell Ackerman. *Handbook of Behavior Problems of the Dog and Cat.* Philadelphia: Elsevier, 2003.

Catanzaro, Tom. *Building the Bond-Centered Practice.* Morrison, CO: Signature Series Monograph, Veterinary Practice Consultants, 2004.

Scott, John Paul, and John L. Fuller. *Genetics and the Social Behaviors of the Dog.* Chicago: University of Chicago Press, 1965.

Horwitz, Debra F., and Jacqueline C. Neilson. *2008 Canine & Feline Behavior.* New York: Blackwell Publishing, 2008.

Classical Conditioning/Pavlov

D. A. Powell, Pavlov, Ivan Petrovich (1849-1936). In: Neil J. Smelser and Paul B. Baltes, Editor(s)-in-Chief, International Encyclopedia of the Social & Behavioral Sciences, Pergamon, Oxford, 2001, Pages 11127–11130, 9780080430768.

(http://www.sciencedirect.com/science/article/B7MRM-4MT09VJ-2C2/2/0bd31f82d90e59c5fd729e80f2d53b66)

Operant Conditioning/Skinner

M. N. Richelle, Skinner, Burrhus Frederick (1904-90). In: Neil J. Smelser and Paul B. Baltes, Editor(s)-in-Chief, International Encyclopedia of the Social & Behavioral Sciences, Pergamon, Oxford, 2001, Pages 14141–14146, 9780080430768.

(http://www.sciencedirect.com/science/article/B7MRM-4MT09VJ-3FF/2/5348e454f6746e6ab9ab4d65271686c1)

APPENDIX A:
PUPPY QUESTIONNAIRES AND CHECKLIST
Puppy Behavior Questionnaire, Version 1

The purpose of these questions is to identify possible areas of concern for clients when dealing with puppy behaviors. From these, you can select which tools to send home with the client. This sample questionnaire identifies behaviors that are addressed by AAHA brochures or the PuppySmarts training DVDs. For more information about how to use this questionnaire, see Step 3: Getting Started under The Step-by-Step Approach section in Chapter 2. If you use other tools, adjust this questionnaire to suit your practice's needs.

A modifiable and printable copy of this questionnaire is available on the companion CD.

Puppy Behavior Questionnaire, Version 1

Questions	Yes (True)	No (False)
My puppy is having problems with housetraining. *PuppySmarts DVD and/or AAHA brochure*	[ ]	[ ]
I would like to learn how to use a crate for my puppy. *PuppySmarts DVD and/or AAHA brochure*	[ ]	[ ]
My puppy nips/bites me or other people. This includes mouthing while playing. *PuppySmarts DVD and/or AAHA brochure*	[ ]	[ ]
My puppy jumps up on people to get attention or to say "hi." *PuppySmarts DVD*	[ ]	[ ]
My puppy chews on things he/she should not. *PuppySmarts DVD and/or AAHA brochure*	[ ]	[ ]
My puppy does one of the following: Hides behind me or barks constantly when strangers approach. Urinates when I reach for him/her or when I first come in to greet him/her.	[ ]	[ ]

	Yes (True)	No (False)

Withdraws and/or puts his/her head down when I reach for him/her or when strangers come near.

AAHA brochure

My puppy demands attention and seems "pushy" with people. [ ] [ ]

AAHA brochure

My puppy seems to bark more often than necessary. [ ] [ ]

AAHA brochure

My puppy likes to dig when outside. [ ] [ ]

AAHA brochure

My puppy has not been regularly exposed to new people, places, animals, and things. [ ] [ ]

AAHA brochure

I would like advice on properly training my puppy. [ ] [ ]

PuppySmarts DVD and/or AAHA brochure

My puppy does not come when I call. [ ] [ ]

PuppySmarts DVD and/or AAHA brochure

I want my puppy to understand *sit, down,* and *stay.* [ ] [ ]

PuppySmarts DVD and/or AAHA brochure

My puppy does not pay attention to me. [ ] [ ]

PuppySmarts DVD

Include additional behavior concerns you may have.

Puppy Behavior Questionnaire, Version 2

This questionnaire is a more detailed version of the Puppy Behavior Questionnaire, version 1. The purpose of these questions is to offer a much broader opportunity to identify possible areas of concern for the client when dealing with her puppy's behavior. From this form, the doctor will select which behaviors the PBA should address in the PBS.

Review the Patient Behavior Questionnaire, and highlight the areas that you will be using. You will also want to review the applicable Puppy Behavior Challenges from Chapter 4 and the relevant sample client handout(s) in Appendix D prior to your appointment, to get a clear understanding of the behavior you will be addressing in the session.

Remember to address only one or two behaviors at most in any one session. More than that could overwhelm the client and reduce your effectiveness. Additional sessions may be necessary to address all of the client's concerns. Always be sure to follow up on the client and patient's progress. If additional sessions are needed, invite the veterinarian into the PBS to prescribe the additional sessions.

A modifiable and printable copy of this questionnaire is available on the companion CD.

Puppy Behavior Questionnaire, Version 2

Questions	Yes (True)	No (False)
My puppy nips/bites me or other people. This includes mouthing while playing. *Refer to Puppy Behavior Challenges (Chapter 4) and the client handout on Biting and Nipping (Appendix D).*	**[ ]**	**[ ]**
My puppy demands attention and seems "pushy" with people. *Refer to Puppy Behavior Challenges (Chapter 4) and the client handouts on Aggressive/Bully Puppies (Appendix D) and Consistency in the Family (Appendix C).*	**[ ]**	**[ ]**
My puppy chews on things he/she should not. *Refer to Puppy Behavior Challenges (Chapter 4) and the client handout on Chewing (Appendix D).*	**[ ]**	**[ ]**
My puppy sometimes likes to get things off my countertops or get into the garbage. *Refer to Puppy Behavior Challenges (Chapter 4) and the client handouts on Drop-it and Grabbing or Training Leave-It (Appendix D).*	**[ ]**	**[ ]**

	Yes (True)	No (False)
My puppy seems to bark more often than necessary.	[ ]	[ ]

Refer to Puppy Behavior Challenges (Chapter 4) and the client handouts on Barking and/or Socialization (Appendix D).

My puppy seems full of energy and has trouble calming down.   [ ]   [ ]

Refer to Puppy Behavior Challenges (Chapter 4) and the client handouts on Consistency in the Family (Appendix C) and Aggressive/Bully Puppies, Grabbing or Training Leave-It, and Biting and Nipping, (Appendix D). This behavior can manifest itself in many ways.

My puppy does one of the following: [ ]   [ ]

    Stops eating and stares when someone approaches him/her while eating.

    Growls when someone approaches him/her while eating.

    Puts his/her ears back and lowers his/her head when someone approaches him/her while eating.

Refer to Puppy Behavior Challenges with special attention to the Food Guarding portion of the Aggressive/Bully Puppies section (Chapter 4) and the client handout on Aggressive/Bully Puppies (Appendix D).

My puppy will sometimes grab onto people's clothes or shoes.   [ ]   [ ]

Refer to Puppy Behavior Challenges (Chapter 4) and/or the client handouts on Biting and Nipping, Grabbing or Training Leave-It, and Hyperactive Puppies (Appendix D).

My puppy jumps up on people to get attention or to say "hi."   [ ]   [ ]

Refer to Puppy Behavior Challenges (Chapter 4) and the client handout on Jumping (Appendix D).

My puppy pants, paces, hides, or cries when I am getting ready to leave the house.   [ ]   [ ]

Refer to Puppy Behavior Challenges (Chapter 4) and the client handouts on Home Alone and Socialization (Appendix D), and discuss with the veterinarian prior to the PBS.

	Yes (True)	No (False)

My puppy's chest is usually wet when I come back home after leaving him/her alone.   **[ ]**   **[ ]**

Refer to Puppy Behavior Challenges (Chapter 4) and the client handouts on Home Alone and Socialization (Appendix D), and discuss with the veterinarian prior to the PBS.

My puppy seems shy, timid, fearful, and maybe even aggressive when meeting new people and animals or is in a new surrounding.   **[ ]**   **[ ]**

Refer to Puppy Behavior Challenges (Chapter 4) and the client handout on Shy, Timid, and Fearful Puppies (Appendix D), as well as the section on Puppy Training Basics in Chapter 4.

When playing with a favorite toy, my puppy does one or more of the following: **[ ]**   **[ ]**

Stops playing when someone approaches him/her.

Starts to growl when someone approaches him/her.

Puts his/her ears back and head down when someone approaches.

Refer to Puppy Behavior Challenges with special attention to the Toy Guarding portion of the Aggressive/Bully Puppies section (Chapter 4) and the client handout on Aggressive/Bully Puppies (Appendix D).

When being groomed, my puppy does one or more of the following: **[ ]**   **[ ]**

Tries to bite me.

Freezes in place.

Pins his/her ears back.

Growls.

Licks me excessively.

Pulls away from me.

Refer to the sections on Grooming and Puppy Training Basics in Chapter 4, as well as the section Bonding Moments in Chapter 3.

Yes (True)   No (False)

When I reach for my puppy, he/she does one or more of the following: **[ ]   [ ]**

    Runs away.

    Tries to bite me.

    Freezes in place.

    Pins his/her ears back.

    Growls.

    Licks me excessively.

    Pulls away from me.

Refer to Puppy Behavior Challenges and Puppy Training Basics in Chapter 4 and/or the client handout on Shy, Timid, and Fearful Puppies (Appendix D) as well as the section on Alternative Methods in Chapter 3.

My puppy is having problems with housetraining.   **[ ]   [ ]**

Refer to Puppy Behavior Challenges (Chapter 4) and/or the client handout on Housetraining (Appendix D).

My puppy sometimes soils in his/her crate.   **[ ]   [ ]**

Refer to the client handout on Housetraining (Appendix D).

I have children at home, and they are becoming afraid of the puppy.   **[ ]   [ ]**

Refer to Puppy Behavior Challenges (Chapter 4) and the client handouts on Hyperactive Puppies, Biting and Nipping, and Grabbing or Training Leave-It (Appendix D). Any or all may be appropriate.

My puppy has not been regularly exposed to new people, places, animals, and things.   **[ ]   [ ]**

Refer to the section on Puppy Training Basics in Chapter 4 and the AAHA brochure.

I would like advice on properly training my puppy.   **[ ]   [ ]**

Refer to the client handouts on Training Methods and Training Principles, especially Turning Lures into Rewards (Appendix C).

	Yes (True)	No (False)

I want my puppy to pay attention to me when I call his/her name.　　**[ ]**　　**[ ]**

Refer to the client handout on Pay Attention (Appendix D).

My puppy does not seem to listen to me.　　**[ ]**　　**[ ]**

Refer to the client handout on Pay Attention and then consider the client handouts on Come, Sit, Stay, Down, Drop-it, and Grabbing or Training Leave-It (Appendix D). Do not give the client all of these forms. Instead, find out where his/her biggest concern lies and start there.

Members of my family sometimes do not agree on how to train our puppy.　　**[ ]**　　**[ ]**

Refer to the client handout on Consistency in the Family (Appendix C).

My puppy resists having his/her collar or leash put on.　　**[ ]**　　**[ ]**

Refer to the section on Puppy Training Basics in Chapter 4, and the client handouts on Collars and Leashes (Appendix D).

My puppy does not like to walk on a leash.　　**[ ]**　　**[ ]**

Refer to the section on Puppy Training Basics in Chapter 4, and the client handout on Leashes (Appendix D).

My puppy pulls when walking on a leash.　　**[ ]**　　**[ ]**

Refer to Puppy Behavior Challenges (Chapter 4), the client handout on Leash Pulling (Appendix D), and the section on Puppy Training Basics in Chapter 4.

My puppy bolts out the door when it is opened.　　**[ ]**　　**[ ]**

Refer to Puppy Behavior Challenges (Chapter 4) and the client handout on Bolting Out Doors (Appendix D).

My puppy will not let go of items he/she has in his/her mouth.　　**[ ]**　　**[ ]**

Refer to Puppy Behavior Challenges (Chapter 4) and the client handout on Grabbing or Training Leave-it (Appendix D). You may also want to review the client handout on Biting and Nipping (Appendix D).

	Yes (True)	No (False)
Has anything recently changed in your household? This could be new additions to the family, children leaving home, new pets, or other changes.	[ ]	[ ]

Sudden changes in the environment may upset the puppy and cause behavior changes. Use Touch Work and T-Shirts, as discussed in the section Alternative Methods in Chapter 3.

	Yes (True)	No (False)
Have you noticed any changes in your dog's behavior?	[ ]	[ ]

Alerts you to something new that previously was not an issue.

Include additional behavior concerns you may have.

Puppy Behavior Session Checklist

Besides the basic questions on this checklist, you may need to probe further to find out why the puppy is presenting a specific behavior. Use open-ended questions such as who, what, when, where, why, and how.

Puppy Behavior Session Checklist

General Questions (always ask in the first session)

These are very important questions. Most clients have no idea what has occurred in their puppy's life before the moment they met. Probe for as much information about the puppy's prior environment as you can. Refer to the section on Understanding the Puppy's History in Chapter 4 for the possible causes of the current behaviors. This will also give you some ideas on what the puppy may need and how to adjust client expectations. Many puppy behavior problems are caused by a lack of social interaction, exercise, and playtime with the family. Boredom can be a big factor in puppy behavior.

❑ Where did you get [puppy's name]?

How did the client hear about the puppy? Pet store? Backyard breeder? Commercial breeder? Gift from a friend? Stray? Shelter puppy?

❑ How old was [puppy's name] when he/she came to live with you?

Older puppies may not have had emotional or social needs met or may have been pushed too hard for many reasons, including the showring.

❑ How much exercise and playtime does your puppy receive?

This question opens the door to a discussion of the puppy's needs. The more exercise and appropriate playtime a puppy receives, the easier he/she will be to train. It will also help you set realistic expectations with the client.

❑ Do other people live in the house with you and [puppy's name]?

Consistency is important, along with making sure clear guidelines for everyone are established. If the answer to this question is "Yes," ask the questions in the next section, Consistency in the Family; if "No," skip to the behavior questions.

Consistency in the Family

It is all about consistency and not sending confusing messages to the puppy. You can read additional information about this in the client handout on Consistency in the Family in Appendix C. You will also need to schedule a PBS with the entire family (or other people living in the home) to go over the puppy's needs, discuss how puppies learn, and to help the family set guidelines.

If the children are old enough, you want them involved with training the puppy, as children are often excellent trainers.

❑ Has the family sat down and written out a plan on how [puppy's name] will be trained?

❑ How old are the children?

❑ Does everyone in the family work with [puppy's name] in the same way?

❑ How is [puppy's name] responding to that?

❑ In what ways are family members working differently with the puppy?

Aggressive and Bully Puppies and Toy and Food Guarding

In this section, you are trying to determine the severity of the issue. Find out if the client has noticed certain people or situations that may aggravate this behavior. We are also probing for how the client is dealing with it and what mistakes she could be making accidentally. Read Puppy Behavior Challenges (Chapter 4) and the client handout on Aggressive/Bully Puppies in Appendix D. If the challenge is with food guarding or toy guarding, you will find this information under the same Puppy Behavior Challenges section on Aggressive/Bully Puppies.

Be aware that often aggressive puppies are actually shy, timid, and fearful puppies. If that is the case, then review the relevant Puppy Behavior Challenge and client handout. The label "aggressive" is sometimes overused and inappropriate, but if that truly is the problem, then perhaps a referral to a board-certified veterinary behaviorist or applied animal behaviorist is in order. When in doubt, speak to the patient's veterinarian on how to best proceed with this patient.

❑ When does [puppy's name] exhibit this behavior?

❑ Could you explain to me what this behavior looks like?

❑ When did you first notice [puppy's name] acting like this?

❑ Does [puppy's name] guard food or toys?

❑ What have you done to address this behavior?

❑ How did that work for you?

Barking

In this section, you are trying to find out whether the client has noticed if certain animals, people, or situations may aggravate the barking behavior. You are also probing for how the client is dealing with the barking and what mistakes he/she could be making accidentally. Read the Puppy Behavior Challenges (Chapter 4) and client handout (Appendix D) on Barking to first understand some of the reasons that could be causing the behavior. You will then be in a position to determine whether this behavior is really excessive or is, in fact, normal. The section on Puppy Training Basics in Chapter 4 may also be helpful when working with this client.

Some clients need help understanding what is normal barking and what is excessive. Helping them to better understand normal behaviors makes life easier for both client and patient.

- ❑ What does [puppy's name] bark at?

- ❑ Does [puppy's name] do that most of the time?

- ❑ When [puppy's name] starts barking, how long before he/she stops?

- ❑ How long has this been going on?

- ❑ Have you had any time to start socializing [puppy's name]?

- ❑ (For small breeds) Do you have a tendency to carry [puppy's name] around?

- ❑ What have you done to address [puppy's name]'s barking?

- ❑ Do you feel like it is working?

- ❑ Have you ever tried praising [puppy's name] when he does quiet down?

Biting

In this section, you are trying to determine the severity of the biting issue. Find out whether the client has noticed if certain people or situations instigate the biting issue. You are also probing for how the client is dealing with it and what mistakes he/she could be making accidentally. Is the biting more than normal puppy biting or is it aggressive? Would a referral to a behavior professional be appropriate? Read the Puppy Behavior Challenges (Chapter 4) and client handout (Appendix D) on Biting and Nipping. When working with this behavior, clients sometimes do not see this as the puppy's way of asking politely to be put down or to be left alone. The puppy may show concern by licking the client's hand, asking the client to politely stop what he/she is doing. Share this information with the client and send the client home with information on how to work with biting via the PuppySmarts training video on Biting, the AAHA brochure on Biting, and the client handout in Appendix D. Look for basic mistakes the client may be making when trying to address this behavior. Without realizing it, the client could be making matters worse.

When appropriate, demonstrate with the client how he/she can work with this behavior at home, and then give the client the opportunity to show you what he/she has learned so the client will be working with this behavior properly at home.

❏ Has [puppy's name] ever bitten you or anyone else hard enough to draw blood?

❏ Can you tell me what was going on when [puppy's name] bit [you/someone else]?

❏ How did you address that with [puppy's name]?

❏ Has he/she bitten anyone else since you originally addressed this with him/her?

Chewing

In this section, you are trying to determine the severity of the chewing issue. You are also probing for how the client is dealing with it and what mistakes he/she could be making accidentally. Read the Puppy Behavior Challenge (Chapter 4) and client handout (Appendix D) on Drop-it. Consider having the veterinarian prescribe the PuppySmarts Chewing video be sent home with the client at the end of the PBS. Some clients may get confused between the words "chewing" and "biting." If biting is the real issue, then the Puppy Behavior Challenge (Chapter 4) and client handout (Appendix D) on Biting and Nipping or the PuppySmarts video on Biting may be more appropriate.

Is the puppy in danger with what he is chewing on? If so, the client should be taught how to properly remove items from the puppy's mouth. Is the puppy simply teething, and does he just need better chew toys to help with new teeth? Does the puppy have a dental problem, such as retained deciduous teeth? When in doubt, refer to protocols established by your veterinarians or excuse yourself from the room to consult with a doctor on how to best proceed.

If the client says, "My puppy is chewing on my hands," then you may need to decide which behavior you will be addressing today. Take the lead with your patients when these challenges arise. If the puppy is biting your hands, then discuss biting. Remember, your first responsibility is to the patient.

When appropriate, demonstrate with the client how he/she can work with this behavior at home, and then give the client the opportunity to show you what he/she has learned so the client will be working with this behavior properly at home.

❏ What kind of objects is [puppy's name] chewing on?

- ❑ When you catch [puppy's name] chewing on something he/she should not be, how do you address it?

- ❑ How has [puppy's name] responded so far?

- ❑ What kind of chew toys does [puppy's name] currently have to chew on?

- ❑ Does [puppy's name] have any hard rubber toys to chew on?

- ❑ Have you thought about teaching [puppy's name] the cue *leave-it*?

Collars and Leashes

In this section, you are finding out what the client has tried and how the puppy responded. Read the sections on Puppy Training Basics in Chapter 4, with special focus on the subheadings Collars and Leashes.

This may be a PBS in which you are showing and instructing the client how to help her puppy become comfortable wearing a collar and/or walking on a leash.

- ❑ How did [puppy's name] react when you tried to put the collar or leash on him/her?

- ❑ What have you done to address this behavior?

- ❑ How has that worked for you?

Crate Soiling

Size does matter. If the crate is too large for the puppy, then the crate may need to be changed. Review the client handout on Housetraining (Appendix D). Offer clear guidelines to the client on common mistakes such as confining the puppy too long or the crate being too large. Give the client some ideas on how to address this issue based on the information on Crate Soiling in the Puppy Behavior Challenges in Chapter 4.

- ❑ How large is [puppy's name]'s crate?

- ❑ How much time does [puppy's name] spend in the crate?

- ❑ When did [puppy's name] start soiling the crate?

- ❑ How have you addressed it so far?

Counter Surfing

In this section, you are trying to determine the severity of the issue. Find out whether the client has noticed if anything triggers or aggravates this behavior. You are also probing for how the client is dealing with it and what mistakes he/she could be making accidentally. Read the Puppy Behavior Challenges (Chapter 4) and client handout on Counter Surfing (Appendix D).

❑   How have you attempted to stop this behavior?

❑   How did that work for you?

❑   Do you put [puppy's name] in the crate when you cannot supervise him/her?

Grabbing

In this section, you are trying to determine the severity of the grabbing issue. Find out whether the client has noticed if certain people or situations instigate this behavior. Is the client making matters worse or better? Is this behavior with food only, toys only, or with many things? Read the Puppy Behavior Challenges (Chapter 4) and client handouts on Grabbing or Training Leave-It and Counter Surfing (Appendix D). You may also want to watch the PuppySmarts Chewing video, which teaches Leave-it. All may prove helpful based on how the client answers the questions below.

When appropriate, demonstrate with the client how he/she can work with this behavior at home. Then give the client the opportunity to show you what he/she has learned so the client will be working with this behavior properly at home.

❑   When [puppy's name] takes a treat from you, does he/she take it gently or grab it and almost take your fingers with the treat?

❑   Does [puppy's name] grab food away from others?

❑   Does [puppy's name] grab articles off tables or countertops?

❑   Does [puppy's name] grab clothing?

❑   What have you tried to address this behavior?

Grooming

In this section, you are trying to determine how the client is handling the puppy's grooming needs. Read the Puppy Behavior Challenges under the subheading Grooming.

Demonstrate, when appropriate, how to trim nails or desensitize the puppy to receiving nail trims, brushing, combing, and dental care. Remember not to push the puppy too quickly. If the puppy is concerned, baby steps will do just fine. You always want to build on the puppy's success.

❑ What grooming tools do you use?

❑ How does your puppy react when you try to brush/comb him/her?

❑ What do you do when that happens?

❑ What happens when you try to clean [puppy's name]'s ears?

❑ What products are you using?

❑ How are you addressing this behavior?

❑ Do you trim [puppy's name]'s nails?

❑ How does [puppy's name] react to you working on his paws?

❑ What do you do when that happens?

Home Alone

In this section, you are trying to determine the severity and possible cause of the separation issue. Find out whether the client has noticed anything that may trigger or aggravate this behavior. How is the client dealing with this behavior, and is he/she encouraging it without realizing it? Read the Puppy Behavior Challenges (Chapter 4) and client handout called Home Alone (Appendix D).

The client's veterinarian should diagnose whether this is a case of separation anxiety prior to your PBS. The doctor may want you to help the client work with this behavior while at the same time prescribing a medication to help the animal.

❑ When did [puppy's name] first start this behavior?

❑ Can you tell me exactly what [puppy's name] does when distressed?

❑ Have you tried anything to stop this behavior from happening?

❑ How has [puppy's name] responded to that?

❑ When you see [puppy's name] starting to act concerned when you are getting ready to leave the house, can you tell me what thoughts are going through your mind?

❑ What is your routine when you are leaving the house?

❑ Do you say anything to [puppy's name] when you are leaving?

❑ How does [puppy's name] respond?

❑ Are you familiar with the term "desensitization"?

❑ What kind of socialization activities has [puppy's name] had so far?

Housetraining

In this section, you are trying to determine the severity of the housetraining issue. You are also probing for how the client is housetraining and what mistakes he/she could be making accidentally. Read the Puppy Behavior Challenges (Chapter 4) and/or client handouts on Housetraining and Crate Training (Appendix D). In addition, depending on the family structure, you may want to read the client handout Consistency in the Family (Appendix C). Listen to the current steps the family is taking and try to identify where the problem may be stemming from—if there really is a problem. Sometimes client expectations can be unrealistic, and the client may need a better understanding of how long housetraining should take. In a perfect world, a puppy can be completely housetrained in three to six months, depending on the animal. Remember that frequent urination could mean an infection. If you believe this may be the cause based on the frequency of the puppy's need to urinate, excuse yourself from the PBS and consult with the patient's or another veterinarian in the practice. Be sure to follow the protocols established by the doctors on how to best proceed with the appointment.

❑ How is [puppy's name] doing with housetraining?

❑ Is [puppy's name] still having accidents in the house? How often?

❑ How do you address the house soiling accidents?

❑ Does [puppy's name] try to hide so you cannot see when he/she eliminates inside?

❑ What do you do when [puppy's name] eliminates in the right area?

❑ How often do you take [puppy's name] outside to relieve himself/herself?

❑ What do you do while [puppy's name] is outside?

❑ Is [puppy's name] on a feeding schedule?

❑ Do you have someone that could let [puppy's name] out in the afternoon while you are at work?

❑ Have you thought of putting [puppy's name] in puppy day care until he/she has better control over his/her bodily functions?

❑ Are you using a crate to confine [puppy's name] when you cannot watch him/her?

❑ Does [puppy's name] ever soil in his/her crate?

Hyperactive Puppies

In this section, you are trying to determine the severity of the hyperactivity. Find out whether the client has noticed if certain people or situations excite the puppy. You are also probing for how the client is dealing with it and what mistakes he/she could be making accidentally. Read the Puppy Behavior Challenges (Chapter 4) and client handout on Hyperactive Puppies (Appendix D). Hyperactivity may also be caused by lack of consistency in the family. Review the client's answers to questions 18 and 24 in the Puppy Behavior Questionnaire, version 2, that the client filled out at the last visit.

When appropriate, demonstrate with the client how he/she can work with this behavior at home, and then give the client the opportunity to show you what he/she has learned so the client will be working with this behavior properly at home.

❑ Is [puppy's name] this active frequently?

❑ Is there something or someone that you believe may be causing this behavior?

❑ Does [puppy's name] demand attention from everyone, or any person in particular?

❑ How have you tried to address his/her demanding behavior so far?

❑ How has that worked for you?

❑ Have you ever praised [puppy's name] in a soft voice while he/she is lying down quietly?

❑ Does [puppy's name] listen to you when you give cues or commands?

Jumping

In this section, you are trying to determine the severity of the jumping issue. Find out whether the client has noticed if certain people or situations instigate the jumping behavior. You are also probing for how the client is dealing with it and what mistakes he/she could be making accidentally. Read the Puppy Behavior Challenges (Chapter 4) and client handouts on Jumping and Hyperactive Puppies (Appendix D). When appropriate, demonstrate with the client how he/ she can work with this behavior at home. Give the client the opportunity to show you what he/she has learned so the client will be working with this behavior properly at home.

❑ Does [puppy's name] jump on anyone?

❑ Could you explain to me what this behavior looks like?

❑ How are you working with [puppy's name] jumping?

❑ How is that working for you?

Leash Pulling

In this section, you are trying to determine the severity of the leash-pulling issue. Find out whether the client has noticed if certain animals, people, or situations aggravate this behavior. You are also probing for how the client is dealing with it and what mistakes he/she could be making accidentally. This can also be a case of lack of exercise. Read the Puppy Behavior Challenges on Leash Pulling (Chapter 4) and possibly the client handout on Exercise and Play (Appendix D).

When appropriate, recommend training tools. If adjustments are needed, do them while in the PBS. Make sure to demonstrate the tool being suggested and always give the client an opportunity to show you what he/she learned from your demonstration so the client will be practicing for success at home.

❑   Does [puppy's name] pull on the leash most of the time?

❑   How often does [puppy's name] go for walks?

❑   Does [puppy's name] pull on the leash no matter who takes him/her for a walk?

❑   In what ways have you worked with [puppy's name] so far to train him/her not to pull on the leash?

❑   Do you ever praise or reward [puppy's name] when he/she walks on a loose leash?

❑   Have you tried any special collars to address this leash pulling?

❑   What effects has this had on the behavior?

Shy, Timid, and Fearful Puppies

In this section, you are trying to determine the severity of this issue. Find out whether the client has noticed if certain people or situations may especially make the puppy fearful. You are also probing for how the client is dealing with it and what mistakes he/she could be making accidentally. Read the sections on Puppy Training Basics and Puppy Behavior Challenges (Chapter 4) and the client handout on Shy, Timid, and Fearful Puppies (Appendix D).

Look for little things the client may be doing to add to the problem, such as carrying the puppy too frequently. Brave behavior by this puppy should always be rewarded.

❑   Is [puppy's name] always this shy/timid/fearful?

❑   When did you first realize [puppy's name] was concerned with things?

❑   Does [puppy's name] act this way most of the time?

❑ Does [puppy's name] act this way with everyone?

❑ Does this behavior occur with strangers?

❑ Does this behavior occur when [puppy's name] is introduced to new environments?

❑ Does [puppy's name] act like this around other dogs?

❑ Does this behavior occur around children?

❑ Does [puppy's name] normally walk on a leash when outside, or do you frequently carry him/her?

❑ How often do you carry [puppy's name] when outside?

❑ Have you tried working with [puppy's name] with these behaviors?

❑ In what ways have you worked with [puppy's name] so far to build confidence?

❑ Have you considered enrolling [puppy's name] in a puppy socialization or kindergarten class?

❑ Do you praise [puppy's name] when he/she is doing something he/she would normally be concerned about?

APPENDIX B:
KITTEN QUESTIONNAIRES AND CHECKLIST

Kitten Behavior Questionnaire, Version 1

The purpose of these questions is to identify possible areas of concern for the client when dealing with his/her kitten's behavior. From these, you can select which tools to send home with the client. This sample questionnaire identifies behaviors that are addressed by AAHA brochures and the client handouts in this book. For more information about how to use this questionnaire, see Step 3: Basic Behavior Program in the section titled The Step-by-Step Approach in Chapter 2. If you use other tools, adjust this questionnaire to suit your practice's needs.

A modifiable and printable copy of this questionnaire is available on the companion CD.

Kitten Behavior Questionnaire, Version 1

Questions	Yes (True)	No (False)
My kitten is having problems consistently using the litter box. *AAHA brochure*	[ ]	[ ]
My kitten sometimes attacks me when playing or when I am sleeping. *AAHA brochure*	[ ]	[ ]
My kitten scratches or chews on things she/he should not. *AAHA brochure*	[ ]	[ ]
Has anything recently changed in your household? This could include new additions to the family, children leaving home, new pets, or other changes.	[ ]	[ ]
Have you noticed any changes in your kitten's behavior?	[ ]	[ ]

Include additional behavior concerns you may have.

Kitten Behavior Questionnaire, Version 2

This version of the questionnaire is more in depth and contains questions referring to the kitten behavior overviews and client handouts. This version is what you would use for Step 5: Advanced Behavior Program in the section titled The Step-by-Step Approach in Chapter 2. If you use other tools, adjust this questionnaire to suit your practice's needs.

A modifiable and printable copy of this questionnaire is available on the companion CD.

Kitten Behavior Questionnaire, Version 2

Questions	Yes (True)	No (False)
My kitten is having problems consistently using the litter box. *Client handout and/or AAHA brochure*	[ ]	[ ]
My kitten scratches or chews on things she/he should not. *Client handout*	[ ]	[ ]
My kitten is a member of our family. *Client handout and/or AAHA brochure*	[ ]	[ ]
My kitten bites me or others. *Client handout and/or AAHA brochure*	[ ]	[ ]
My kitten scratches me and/or others. *Client handout*	[ ]	[ ]
I sometimes play with my kitten using my hands or feet. *Client handout*	[ ]	[ ]
My kitten likes to jump on countertops and tables. *Client handout*	[ ]	[ ]
My kitten likes to knock things over in the house. *Client handout*	[ ]	[ ]

	Yes (True)	No (False)
My kitten pounces on my legs and/or feet.	[ ]	[ ]
Client handout		
Have you noticed any changes in your kitten's behavior?	[ ]	[ ]
Has anything changed in your household that may be affecting your kitten?	[ ]	[ ]
If you have other pets in your home, are they getting along with your kitten?	[ ]	[ ]
Client handout		
Is bath time for your kitten a challenge?	[ ]	[ ]
Client handout		

Include additional behavior concerns you may have.

Kitten Behavior Session Checklist

In the general section of this checklist are questions about where the kitten came from and how the client acquired the kitten. This will let you know if the kitten came from a reputable breeder, a pet store, a shelter, a backyard breeder, or if the family found the kitten dumped off somewhere or the kitten found them. Each scenario may present unexpected challenges.

Besides the basic questions on this checklist, you may need to probe further to find out why the kitten is presenting a specific behavior. Use open-ended questions such as who, what, when, where, why, and how.

Kitten Behavior Session Checklist

General Questions (always ask in the first session)

These are very important questions. Most clients have no idea what has occurred in that kitten's life before the moment they met. Probe for as much information about the kitten's prior environment as you can. The kitten's prior background can reveal possible causes of the current behaviors. These questions will also give you some ideas on what the kitten may need and how to adjust client expectations. Many kitten behavior problems are caused by a lack of social interaction and playtime with the family. Boredom can be a big factor in cat behavior.

❑ Where did you get [kitten's name] from?

Pet store? Backyard breeder? Commercial breeder? Gift from a friend? Stray? Shelter?

❑ How old was [kitten's name] when she/he came to live with you?

Older kittens may not have had their socialization or developmental needs met.

❑ Do you have other pets in the home?

The interactions between the pets could be a factor in the kitten's behavior.

❑ How much exercise and playtime does your kitten receive?

This question opens the door to a discussion on the kitten's needs and setting the client's expectations. It also allows you to ask how the family plays with their kitten. Remember, hands and feet are not play toys for kittens. Cat toys are the only toys kittens should be taught to play with.

❑ Do other people live in the house with you and [kitten's name]?

Consistency is important, along with making sure clear guidelines for everyone are established. If the answer to this question is "Yes," ask the questions in the next section, Consistency in the Family; if "No," skip to the behavior questions.

Consistency in the Family

It's all about consistency and not sending confusing messages to the kitten. You can read additional information about this in the client handout Consistency in the Family in Appendix C and the client handout Introducing Your Kitten to Your Home in Appendix E. You will also need to schedule a family appointment to go over the kitten's needs and help the family to set guidelines.

If the children are old enough, you want them involved with training their kitten, as children are often excellent trainers.

❑ Has the family sat down and written out a plan on how [kitten's name] will be trained?

❑ How old are the children?

❑ What are each child's responsibilities for the kitten?

❑ Does everyone in the family work with [kitten's name] the same way?

❑ How is [kitten's name] responding to that?

❑ In what ways are family members working differently with the kitten?

❑ How is [kitten's name] and the other animals getting along?

Counter Surfing

You are trying to find out if the kitten has an alternative to being up on the counters or if there is something on the counters that is attracting the kitten. Kittens and cats love to be in high places to watch their surroundings. Refer to the Kitten Behavior Overviews (Chapter 5) and client handouts on Counter and Table Surfing (Appendix E).

❑ When did [kitten's name] start jumping on things?

❑ Were there interesting things for [kitten's name] to explore?

❑ Have you removed those items yet?

❑ How has [kitten's name] responded to that?

❑ Does [kitten's name] have a cat tree or cat climber?

Knocking Things Over

Kittens are very curious and can get bored easily. You are trying to find out if the kitten has enough stimulation in his/her environment. Refer to the Kitten Behavior Overviews (Chapter 5) and the client handout on Knocking Things Over (Appendix E).

- How often does [kitten's name] knock things over?

- Are there certain items [kitten's name] seems to knock over more than others?

- What kind of things does the kitten have to smell, taste, or play with?

- Does your kitten act bored?

- Does this behavior happen while you are home or only when you are away?

- Why do you think [kitten's name] might be doing this?

- How have you addressed this behavior so far?

- How did [kitten's name] respond to that?

Litter Box Training

In this section, you are trying to determine the severity of the litter box training issue. We are also probing for how the client is litter box training and what mistakes could he/she be making accidentally. Review the section Kitten Behavior Overviews (Chapter 5) and the client handout Litter Box Troubleshooting (Appendix E).

- How is [kitten's name] doing with housetraining?

- When did you notice [kitten's name] stop using the litter box?

- Has anything changed in your home since acquiring your kitten?

- Has [kitten's name] ever been consistent at using the litter box?

- If so, why do you think [kitten's name] stopped?

- Have you changed the location of the litter box?

- Have you changed to a different brand of litter?

- How big is [kitten's name] litter box?

- How often do you change the litter?

- How often do you wash out the litter box completely?

- What household cleaners do you use to clean the litter box?

- Where is the litter box located in your home?

- Does this location offer [kitten's name] privacy?

❑ Does it offer [kitten's name] an escape route?

❑ Would you say the litter box is located in a high-traffic area in your home?

❑ How have you tried to address this behavior so far?

❑ How did [kitten's name] react to that?

Pouncing on Legs

Kittens can get bored easily. When they do not receive appropriate interactive playtime with the family, they can invent their own games to get the client's attention. Refer to the Kitten Behavior Overviews (Chapter 5) and client handout Pouncing on Legs (Appendix E).

❑ When does [kitten's name] usually pounce on you?

❑ How have you addressed this behavior so far?

❑ How did [kitten's name] respond to that?

Scratching and Biting

You are asking these questions to find out whether the family is playing with the kitten in an appropriate manner. In addition, you want to make sure the kitten has an outlet for his/her scratching needs. Refer to the Kitten Behavior Overviews (Chapter 5) and the client handout Scratching and Biting (Appendix E).

❑ How often does [kitten's name] scratch or bite you?

❑ When does this usually happen?

❑ Has [kitten's name] scratched or bitten anyone else?

❑ How do you play and interact with your kitten?

❑ Is [kitten's name] scratching on furniture or other inappropriate items?

❑ Does [kitten's name] have a scratching post?

❑ How many scratching posts does [kitten's name] have?

❑ Does [kitten's name] use his/her scratching post?

❑ How have you addressed this behavior so far?

❑ How did [kitten's name] respond to that?

APPENDIX C:
SAMPLE CLIENT HANDOUTS (TRAINING PRINCIPLES)

Consistency in the Family

It is important for your entire family to work as a team to successfully train your new pet. Your puppy/kitten will need to learn the meaning of your words, the rules in your home, and your expectations of him or her. You will confuse your puppy/kitten if words, rules, and expectations are inconsistent from one family member to another.

Children can be great little trainers. It is recommended they be at least three years of age to help with training. Children should be supervised during training sessions. Pets and children should never be left alone unsupervised.

Since puppies do like to chew and kittens like to sharpen their claws, discuss these natural behaviors of puppies/kittens with your children before their favorite toys are chewed, clothing is ruined, and various other problems occur. Let your children know that cruelty (actions such as shocking, hitting, shaking, pulling ears, pulling tails, grabbing, or rubbing the pet's nose in feces) will not be tolerated under any circumstances. Many times children don't even realize that what they are doing is cruel to the animal. In addition to children, babies and toddlers should never be left alone with your pet or even with your most trusted adult dog, unsupervised.

Puppies and kittens have some basic needs—physically, mentally, and emotionally. They need to eat on a regular schedule (consult your veterinarian) and require access to clean, fresh water. Puppies need to relieve themselves frequently as well as nap many times throughout the day. Playtime, rest, and exercise are all important to young dogs or cats.

Puppies and kittens should have a safe, comfortable place to go when they are not being supervised. Ideally, puppies/kittens should be socialized a few minutes each day, and they should be trained for 5 to 10 minutes at a time. This time can be gradually increased as the pet gets older. Each training session should remain short to make it easy for the puppy/kitten to pay attention, but more sessions can be added throughout the day.

To care for your pet and to maintain consistency while training, develop an action plan for your family. Adults should be actively involved in supervising the following activities. Consider these questions when developing your plan:

- Who will feed the puppy/kitten and when?
- Who will keep the water bowl clean and filled with fresh water?
- Who will be in charge of the puppy's/kitten's bedding to make sure it is clean?
- What training techniques will you use for housetraining? Crate training? Litter training?

- Who will be in charge of training the puppy/kitten in these areas?
- Who will socialize the puppy/kitten to different people, places, and things? How often and when?
- Who will play with the puppy/kitten? How often and when?
- Who will take the puppy for a walk? How often and when?
- Who will brush the puppy/kitten? How often and when?
- Who will trim the puppy's/kitten's nails?
- Who will brush the puppy's/kitten's teeth? How often and when?
- Who will supervise the puppy when not in his/her crate? How long and when?

Develop a chart outlining everyone's duties. Have each family member check off their chores as they are accomplished daily and continue this for a three-month period.

Give your puppy or kitten room to make mistakes. Being overly demanding and short tempered can have long-term negative consequences on the behavior and enjoyment of your pet over its lifetime.

Always be consistent. If your pet is not allowed to do something today by one person and then allowed to do the same thing tomorrow with someone else, the pet will become confused with the mixed signals. This can even happen with the same person from one day to the next! Mixed signals can confuse your pet and lengthen the time it takes for puppies and kittens to learn what is right, wrong, acceptable, and not acceptable.

Puppies and kittens can express their confusion by being overly excited, fearful, shy, timid, or aggressive. In many cases, you may see a combination of behavior problems occurring.

This does not have to happen. Your pet is constantly learning, and every waking hour to them is a learning experience. With time, patience, and training consistency, your pet will become a successful member of your family. End all training sessions on a positive note. If the puppy/kitten is having difficulty learning a new cue, stop before you both get bored or frustrated and ask the pet to do something he/she is very good at, such as asking him to sit. Mark and reward this behavior and end the training session.

Families that work together as a team can be great trainers. Keep the same message every time (be consistent). If more than one family member is training, all family members should train the exact same way. Introduce variations into your training (different locations, different people, and distractions) gradually.

It does not matter what you are trying to teach your pet, just be consistent. If you are training a puppy to do his business outside, then take the puppy out the same door every time. If you want your puppy to be quiet while in his/her crate, don't open the door to the crate when the puppy is barking to be let out. If you do not want your kitten scratching or biting at your hands, never use your hands to play with your kitten. Use an appropriate cat toy instead. Reward the behaviors you want in a consistent manner.

Some people get upset with an animal because they think the puppy/kitten knows what they want. They call their pet hardheaded, or stubborn, or they say their pet has selective hearing. This can happen when you think the pet knows what the word "sit" means. You taught *sit* in the kitchen to receive meals, and now you are outside and the puppy/kitten just will not listen to you. It is not because the puppy/kitten did not hear you, nor does the pet have selective hearing. It is because your puppy/kitten really does not know what "sit" means outside.

Dogs and cats do not generalize very well—they learn in context. That means the puppy/kitten has to be taught *sit* in many different places for the first few times, and if possible with different family members. Once he/she understands *sit* in various places with different people, the puppy/kitten now understands that "sit" means put its rear end on the floor whenever it hears that word.

Training Methods

There are four basic training methods often used in training animals. Decide which method you want to use with your newest family member. Here is a little information about each method to help you decide.

Verbal Training

Verbal training consists of saying a word such as "yes" whenever your puppy/kitten does what is expected. Coupled with a reward, this reinforces the behavior that you want. Perfect timing is not necessary since the animal can also pick up on the emotional content of the mark ("yes" in a happy voice). For example, if you give the cue "sit," when the puppy's rear hits the floor, you mark it with "yes" (happy voice) and give the puppy a reward.

Target Training

Target training involves an object that your pet learns to touch with his/her nose. Once the pet touches the target from a short distance, he/she can learn to touch the target from a greater distance.

This method of training is great for getting a puppy or kitten from point A to point B without pulling or having to drag the pet into a position. Targeting can also be helpful in training a pet to do tricks.

To begin target training for larger breeds of dogs, use the palm of your hand as a target. Place the palm of your hand in front of the puppy's nose. Say the word "target," and wait for your puppy to lean forward and touch your hand. The second your puppy leans forward and touches the palm of your open hand with his/her nose, mark the behavior (the puppy touched your hand) and give him/her a treat to reinforce the behavior the puppy just gave you.

With smaller-breed puppies and kittens, you can use a targeting stick or an old wooden spoon. Take a piece of colored tape or use a marker and draw a line around the spot on the stick or spoon you want the pet to touch. Place the stick or spoon an inch in front of the animal's nose. The second the puppy/kitten touches the targeting stick or spoon, mark and reward your pet.

If your pet does not reach out to touch the object, try rubbing a little cheese, chicken, or fish on your hand or on the targeting stick. This will usually get the animal's attention. Repeat the above exercise with the new smell on the object.

Once your animal consistently touches the object when you say the word "target," extend the distance between the animal and the target or your hand. Remember, every time you ask the pet to *target*, it is your job to mark and reward when he/she demonstrates the desired behavior.

Continue to gradually extend the distance until the puppy/kitten will walk around in a circle or across the room to *target* the targeting stick or your hand when asked to do so.

Clicker Training

Clicker training is similar to using verbal training, but instead of using your voice, you use your clicker. A small click from the clicker can be used to mark a requested behavior or to form more complicated behaviors. Timing in this method is extremely important. It requires the trainer (you) to click the moment the animal performs the desired behavior; for example, the second your puppy's rear hits the floor when training the *sit* cue.

Before training your animal, train yourself to properly administer the timing of the clicker. Hold a tennis ball in one hand and the clicker in the other. Bounce the tennis ball, and the moment the ball hits the floor, click. Your click should be made at the exact moment the ball hits the floor. You may need to practice a few times to get your timing correct.

If you think you and your pet would enjoy clicker training, then by all means, give it a try. Clickers are available at most pet stores.

Hand Signals

Hand signals can be used as a replacement for verbal training. It is most commonly used when training a deaf animal, when training from a distance, in competitions, or just for fun. There are many different hand signals you can use in training. What they look like can be left up to you unless you are planning to compete in an obedience ring with your pet.

Training Principles

What you allow today, your puppy/kitten will try again tomorrow. In many cases, this will be a good thing.

To the same effect, what you ignore today, your pet will learn to stop doing in the future. Your pet will try the same behavior many times until he/she realizes that the behavior is not going to be rewarded. When the animal makes that connection, the behavior will stop. Puppies, kittens, and children are always testing their boundaries.

Here are two simple rules to follow as your pet is learning the do's and do not's about his/her new home:

- Behaviors that are rewarded will increase in frequency, intensity, and duration.
- Behaviors that are ignored (not rewarded) will diminish in time, frequency, and duration.

If you like a behavior your puppy or kitten is exhibiting, give it a name and then mark and reward it. If you do not like the behavior the pet is displaying, walk away, look away, isolate him, or ignore him; do not scold, reason, or talk to him. Just ignore him when possible and walk away. This will ensure he is not rewarded for poor behavior, positively or negatively. However, do not ignore him or walk away if he is doing something that could hurt him or someone else.

The most important thing your puppy wants is your attention. Kittens are a bit more aloof, but they still enjoy getting attention. As far as your pet is concerned, whether you are scolding, pushing, pulling, or yelling, it is still attention. When you pay attention to your pet with words, pets, toys, or treats, he/she will want to perform more of those behaviors because he/she gets your attention. Making a short, sharp sound with a can or plastic bottle with coins in it, clapping your hands together, or a loud word such as "off" will interrupt the pet's behavior. When the puppy/kitten stops the inappropriate behavior, remember to mark and reward him.

You will want to make some decisions in regard to the training methods everyone in your household will use. Harsh corrections are no longer considered productive when training animals. Besides, your new family member deserves more respect than that.

For any type of training to be successful, everyone must use the same training methods. When different family members use their own style while training, the pet will only get confused and you will be setting your pet up for failure.

Although done differently, every puppy and kitten needs and deserves appropriate playtime and an opportunity to burn off excess energy every day. People who give their animals the exercise and playtime they so desperately need while young will find training will go much faster and easier than for animals who do not receive these opportunities.

Puppies and kittens who receive consistent guidelines and daily opportunities for exercise and play will become wonderful family members. Training, patience, consistency, and love are the keys. Consider your baby pet your newest savings account. What you put in today, you will get back in the future with interest (a well-behaved and trained pet for perhaps the next 10, 15, or 20 years).

Marking the Correct Behavior

Like most animals, one of the ways your dog learns is by associating things that happen at the same time. For reward-based training to work, the reward must be given as soon as the desired behavior is offered. This should happen within a half second from when the puppy offers the desired behavior. This is not always possible, however, unless your puppy is right by your side. To get around this problem, you can create a temporary substitute for the reward that becomes associated with the concept that a reward will come shortly. This learning process is called associative or classical conditioning. The temporary reward substitutes are called a mark (or a bridge).

A mark can be any word or other type of signal, just as long as it is used consistently. For example, you could use the word "yes" or a click from a clicker as a mark. You may choose to use another word, but it must always be the same word. In time, your puppy/kitten will begin to associate that something good is going to happen when he/she hears the word "yes," the click from the clicker, or whatever other word you have decided to constantly use.

Timing and consistency in all training are very important. The mark should happen the moment the desired behavior happens. It is important to follow up the mark quickly with the reinforcement reward (treat, pet, or play).

Turning Lures into Rewards

A lure is something you use to guide your pet into a behavior. A reward is something your pet receives after he/she offers a behavior. The most commonly used lures and rewards are food treats, although toys can be effective.

The lure is used to coax an animal into a behavior you want him/her to do. It physically guides the animal into the desired position, such as a *sit* or a *down*. You can use a lure to entice your pet to *come* when he/she is called by showing him/her the treat.

A reward just seems to "magically" appear when the pet offers a desired behavior. This differs from a lure in that your animal may anticipate a reward, but does not know for sure if there is one or when it will appear. If you have a visible reward in your hand, he will learn to offer the desired behavior only when you have the reward in your hand. In some situations, rewards may be visible, but they are not used as lures. A perfect example would be playing fetch with a dog. The dog sits, and then you throw the ball.

At first the reward may be food treats, as most puppies and kittens are motivated by food; however, the reward can be play, a favorite toy, or an ear scratch. Whatever motivates your animal most will work as a reward. Rewards work only if the reward being used is important to the animal. You will want to take some time to learn what motivates your pet.

When training your animal, you must determine what he/she loves and likes. Training treats are not meant to be meals. Always use tiny pieces of food so that your puppy/kitten does not fill up on them. Training treats should be low-fat and may include small, healthy commercial treats or tiny pieces of cheese, liver, chicken, fish, chicken hot dogs, or beef. Training toys for puppies to be used as rewards can be squeaky toys, tennis balls, a Frisbee, or any toy your puppy really enjoys. Training toys for kittens can be dancing feathers or tassels, catnip, a soft mouse, or other animated objects.

Offering your pet his/her normal mealtime kibble is okay if you are asking for an easy behavior without distractions. In many cases, though, kibble will not be enough to motivate your pet if there are distractions present.

Once your animal understands the cue you are requesting, treats should be used intermittently and unpredictably. One time you may reward with one tiny treat, another time two, then you may give an ear scratch or you may choose to use a favorite toy as the reward. The most important thing you can do is to keep your animal guessing.

When your pet goes from the lure to the reward for the first time or two, it is time to introduce him/her to a jackpot. A jackpot is a super-reward for doing something outstanding, such as many treats given for one behavior. For example, when your animal figures out that the cue *sit* means "butt on floor" without being lured into the position with a treat, that is a big accomplishment, and your pet deserves a big reward. Instead of one or two little treats, give him/her a handful of treats, such as five or six treats, one after the other, until your pet has received all of them.

Sometimes people get confused when training with food rewards. Their concern is that if the animal does not see the reward, they will not give the desired behavior. This is true only when you forget to turn the lure into a reward. See the following two examples.

Example: Using a Lure

Say you want to train your animal to *sit* on cue. Show your pet the lure/treat in your hand while he/she is standing. Place your hand with the treat in it right above his nose and slowly move your hand with the treat in it over his head to a position just a hair above his nose. As you slowly move over the nose toward his head, his head will tilt upward to follow the treat until the only way he can keep an eye on that tasty treat is to put his butt on the floor. You have just lured him/her into a *sit*. Once the pet sits, mark (a word like "yes" or a click from your clicker) and reward him/her for giving you the behavior you requested.

Example: Combining Lures and Rewards to Train a Common Behavior

As in the example above, use a lure to begin training the *sit* command. You will want to begin associating a mark with the reinforcement (treat). When the animal's butt hits the floor, mark the behavior (sitting) with a word such as "yes" or use a clicker, if you prefer. Once you have marked the desired behavior, quickly give the animal the treat. Repeat this exercise 15 to 20 times over three or four days.

Once the animal is performing the intended behavior consistently, it is time to add the word "sit." The reason you wait to introduce the verbal cue is because he/she knows what is expected (sitting). Now is the time to associate a cue with the behavior. When you lure the pet into the behavior, say the word "sit" the second the animal's butt touches the floor. As soon as the behavior is completed, mark and reward your pet. Repeat this exercise seven or eight times a day (for 10-15 seconds) over the next couple of days, or until you are sure your pet is getting the idea; then it is time to replace the lure with a reward.

Keep the treat you are using as a reward out of the animal's sight this time, but have it available quickly when you get the correct behavior. Using your hand the same way (but without a treat), give your pet the verbal cue "sit" and wait for him/her to think through your request. Do not repeat the cue. Give your pet a few seconds to figure out your request. The second his/her butt hits the ground, mark the behavior with a "yes" or click and quickly give the pet a jackpot (many tiny rewards) to reward a great job.

Now that your pet understands the *sit* cue, you no longer need to use the food as a lure to coax him/her into sitting from standing. Instead, you give the verbal cue "sit," butt hits floor, and you immediately mark that *sit* with "yes" or a click from your clicker to let the animal know he/she has done exactly what you wanted, and then you reinforce the correct behavior with a treat as a reward. The reward reinforces the desired behavior.

Rewards Schedule

When you begin training your puppy/kitten, one very important piece of the training is a reward schedule. A reward schedule refers to how many and how often your pet will receive treats from you. The concept of a reward schedule is to keep your animal interested and guessing about what he/she may or may not receive when offering the requested behavior.

This schedule will start off very basic and then vary as the requested tasks become more difficult, other family members begin to train him, and distractions are added to the training schedule.

Good trainers are quick, both in marking the correct behavior and rewarding it. They are generous, yet unpredictable, with their rewards. They are unpredictable by changing the number, type, and even how often they use different rewards. This scheme will keep your animal guessing, make it fun, and keep your pet's attention focused on you.

Training Sessions

Training sessions should be short, lasting only five to ten minutes. By keeping the lessons short, you will be able to work with your young puppy's/kitten's attention span and fit the training into your hectic schedule. Later, as your puppy/kitten gets older and the animal's attention span increases, you can increase the length of the training sessions.

That does not mean, however, that a one-hour puppy or kitten class is not a great idea. In fact, these classes are a wonderful learning experience for both you and your pet.

When you start a training session, it is important to be relaxed and pay attention to your animal. Begin training in a quiet place where there are no distractions. As the animal becomes successful in learning a new behavior and all family members can get the pet to perform the desired behaviors on cue, then you can begin adding distractions.

Distractions

As distractions are added to the training sessions, reward levels should be increased based on the level of distraction. For example, suppose you ask your puppy to *stay* and introduce a ball on the floor. At first, just ask your puppy to *stay* for a few seconds, then reward and release him/her from the cue. Instead of a regular treat, this time your pet might get a tiny piece of chicken! Wow, that really makes staying still worthwhile because he/she just got a great reward!

When training *stay*, gradually increase the length of time you ask the pet to *stay*. Once he/she has learned to *stay* for 60 seconds, you can introduce a distraction. This could be a ball rolling on the floor in front of the pet or an animated toy on the floor nearby. If the pet holds the *stay*, give him/her two, three, or four tiny pieces of meat or fish and then release him/her from the *stay*. This way, as cues become more difficult, the animal is willing to do its best in hopes of receiving that great reward for its efforts.

Treats

In the beginning, every time your puppy/kitten performs a desired behavior, reward him/her with a food treat. In order to give the animal a food treat that he/she will consider a reward, you need to understand what foods your pet likes and is willing to work for. Different treats will obviously have different values for your animal. Cooked chicken, chicken hot dogs, freeze-dried shrimp, or liver would be a better motivator than normal kibble. Learn what motivates your pet. As a general rule, you would use "normal" rewards for normal performance and save the "high-value" rewards for higher levels of performance. Mix them up a little from time to time so that your puppy/kitten never quite knows what reward he/she may get.

You can try offering a piece of kibble as a reward if the pet is hungry, there are no distractions, and you are asking for an easy cue such as *sit*. If kibble does not interest your pet, you may want to find another treat he/she is willing to work for. It may also be helpful to schedule your training sessions before mealtime, so you will know the pet is hungry.

Today, there are many healthy treats on the market to choose from. Look for treats that can be broken into many small pieces and that do not crumble on the floor when you break them up. When you drop tiny crumbs on the floor while breaking up the treats, your pet especially will be more interested in hoovering (playing vacuum) than paying attention to you. Test a variety of treats until you find the ones that motivate your animal. Only use healthy, low-fat treats and never use candy as a reward for your puppy/kitten because many kinds of candy are poisonous to pets.

Once your puppy/kitten is capable of performing a particular behavior on cue, you can start varying the rewards. You may offer one treat when your pet gives you the desired behavior, and the next time offer two. Or, you may give your pet one treat the first time he/she gives you a requested behavior and then an ear scratch the second time. Remember to keep the treats out of sight once you are past the luring stage in training a desired behavior.

Involving the Family

In order for your puppy/kitten to respond to all family members, everyone should take turns in training. This is a great time to introduce different levels of rewards, meaning treats that your animal places a higher value on. For example, an adult who spends a good amount of time with the pet might get by with just pieces of kibble as rewards. A three-year-old might need more interesting treats such as a small piece of chicken or shrimp to help the animal pay attention, so the child can see positive results from his/her training sessions. At the same time, you want the puppy/kitten to stay focused on the child.

Children should train the puppy/kitten the same way adults do, with short lessons, varying the rewards, and always ending a session on a positive note. Children should never be left unsupervised with a puppy/kitten or even an adult animal. This is when so many accidents happen. Constant supervision will ensure the safety of both the child and the animal.

Evolving

As your puppy/kitten's training request becomes more challenging, more interesting rewards should be offered. As time goes on and the animal becomes more proficient at accomplishing his/her tasks, rewards should become more varied and valuable. You can increase or decrease those rewards, depending on the puppy's/kitten's consistency and response time.

Remember to always end a training session on a positive note. If you are asking for a new behavior from your puppy/kitten and he/she is having a difficult time with it, go back to something your pet understands so you both can be successful before you end your session. Make your training sessions *fun* for both you and your animal. Your pet will learn to look forward to each session, and you can both continue to learn and have great fun in the process.

APPENDIX D:
SAMPLE CLIENT HANDOUTS (PUPPIES)

Aggressive/Bully Puppies

If at any time while working with your puppy you feel threatened or concerned, contact our office for a referral to a veterinary behaviorist. Do not try to solve this problem alone. There are wonderful professionals who can help you and your puppy.

There are three main things you want to teach your puppy. First, you and not the puppy control all resources. Second, the puppy must learn self-control. Third, you want to build your pet's confidence and desensitize him/her to the "triggers" that may be causing aggressive or bullying behavior

The following examples of clear guidelines you need to establish between you and your puppy will teach him that you control all resources and help him learn self-control. Be patient because learning self-control is not an easy lesson.

- *Your puppy does not receive meals until he is sitting quietly.* If your puppy jumps up or nips at you, put the food in the refrigerator or cabinet and walk away. Offer it again when your puppy calms down. Go into the kitchen and get the food bowl. Ask your puppy to "sit." If the inappropriate behavior continues, repeat the exercise until the bowl can be put on the floor and your puppy waits until you release him to eat. You can use the cue "free dog" as the release, and let him eat his meal.

- *When you go outside for a walk, ask your puppy to "sit."* You walk through the door first, and then release him to come outside with you.

- *When you come back inside from your walk, ask your puppy to "sit."* You come in the house first, and then release your puppy.

- *You decide when your puppy receives attention, not your puppy.* This can be a big one for many puppies. When your puppy comes over to you and jumps up, turn and walk away. When your puppy nudges your arm asking to be petted, ignore him and walk away. Whatever you do, do not pet your puppy when he is demanding your attention. When your puppy barks at you, ignore him and walk away. These are all attention-seeking behaviors, and your puppy needs to practice self-control. You want him to learn that being quiet, patient, and polite is the only way to get your attention!

- *You decide who gets to lie where and when.* If your puppy is lying down comfortably in the middle of a room or in the hallway, ask him to move so you can pass. If your puppy does not move, tap him lightly with your foot and ask him to move again. Giving your puppy a cue such as "excuse me" would work well with this. Do not step over your dog.

- *You decide where your puppy gets to sleep, not your puppy.*
- *You decide when and where your puppy gets meals, not your puppy.*
- *You decide if and when you are going to groom your puppy, not your puppy.*

When your puppy is being polite, patient, or quiet, it is very important to mark and reward him for the good behavior.

When your puppy is lying down quietly, walk over and gently pet him, and tell him how good he is being. If your puppy growls or snaps at you, ignore him and walk away. Repeat the exercise a little later when your puppy is lying down quietly. Continue trying this until your puppy can accept your attention while lying down. This does not mean that whenever your puppy is resting, you should bother him. Only try this a few times a day until your puppy is comfortable being touched while lying down quietly.

Everyone in the house must work with the puppy the same way, using the same guidelines, or it will confuse him. Many puppies try very hard to do the right thing, but when they get mixed signals from different family members, they can get frustrated. This frustration can show itself as aggression. The puppy simply does not understand what is expected of him. Sometimes this confusion can be done by a single person or it may be many different people. For example, the puppy can get the message that a behavior is okay to do today, but not tomorrow, but he can do it again the following day—maybe. A puppy cannot understand the inconsistencies and sometimes can become overwhelmed trying to understand humans. Consistency will always play a major role in your puppy's success in settling into his new home.

These are all your choices to make and not the puppy's. Keeping things in order for your puppy sends a clear message that you are in control of all resources. Keeping clear guidelines sets the stage for more appropriate behaviors.

Confidence building involves appropriate socialization opportunities for your puppy and desensitizing him to his new world. Socialization is very important for every dog of any age. Offer new experiences slowly at your puppy's pace so you can build on his successes. Confident puppies are rarely bullies or aggressive.

If, after working with your puppy for a few weeks, you feel the above exercises are not working, please contact our office for a referral to a good behaviorist for additional help in working with this behavior.

Barking

Barking is a natural way for dogs to communicate. Sometimes this communication can be excessive and quite annoying. To be effective in changing this behavior, take a little time to understand what triggers your puppy's barking.

Your puppy may be barking for several different reasons, which may include any of the following:

- He has way too much energy and needs more exercise than what he is currently receiving.
- He is bored and needs toys for some mental stimulation.
- He is fearful and needs more socialization and/or desensitizing opportunities.
- He has been carried around too frequently and needs to experience life with all four paws on the floor to build his confidence.
- He is offering a normal alert bark. This bark is to let you know something is different.

Like any behavior, you do not want to accidentally reward him by paying too much attention to his barking. When too much attention is given to any behavior, it will increase in intensity and duration. Since puppies consider any attention to be a reward, you may be rewarding a behavior you are trying to stop, which will increase the behavior. It is important that you do not try to reason with your puppy. Reasoning takes time, words, and most of all, your attention. Your attention is the one thing your puppy will always want from you.

The following are some things to think about and consider doing with your puppy to stop excessive barking.

Exercise

Puppies in general need lots of exercise. If your puppy is not getting enough exercise, then increasing his activities may help rein in his barking behavior. Taking him for a walk is exercise, but not enough for a young dog. If you have a fenced yard, then playing fetch with him outside is a great exercise. If you have a friend who has a safe dog that is up to date on vaccines, you could invite the other dog over for a playtime with your dog in the backyard. If you live in an apartment, a good game of fetch indoors can serve as an exercise opportunity for him as well. Playtime is very important to a young dog. Besides getting exercise, playtime teaches your puppy how to interact with people. Belly rubs, fetch, and puppy-in-the-middle are all games you can play with him while teaching manners at the same time.

Boredom

If your puppy is bored, offer him new things to figure out on his own. You can purchase a few hard rubber toys that you can put some treats in for him to work for. Give your puppy one toy a day to play with until he figures out how to get all the treats out of the toy. Then put that toy away and offer a new toy the following day. Exchange toys frequently so your puppy does not get bored with the same toy. You can fill these toys with a little canned dog food, treat spreads, treats, or peanut butter. Letting your puppy figure out how to get the treats out of the toy is great mental stimulation for him. (*Caution*: Do not use peanut butter if anyone in your home is allergic to peanuts.) Most of these toys can be placed in the dishwasher for cleaning, but check the manufacturer's recommendations first. When puppies have access to all of their toys at the same time, they can quickly become bored with all of them.

Socialization

Many puppies who are excessive barkers lack socialization. They bark excessively because they do not understand enough about the world around them and find it a fearful place. If your puppy needs more socialization, take him to different places with you. Slowly introduce him to different people, places, and things.

Socializing your puppy is one of the most important things you can do for him. If there is a puppy socialization class you can enroll him in, do so. This is a wonderful opportunity for him to be socialized with other people and puppies. Walks in the park or at a playground can also be a great way to socialize your puppy. Make sure any other animals that your puppy meets prior to completing his vaccines have received all of their vaccines as well. When taking your puppy to any outdoor activity, it is important that you protect him. If a stranger comes up and wants to pet him, let the puppy go to the stranger; do not force your puppy to hold still for a pet from anyone.

Notice if your puppy seems more concerned with one person than another. Is he showing more concern to people with glasses, or hats, children, men, or people in uniform? Whatever the stimulus is for him, look for a pattern of his concerns. If you do find something he seems consistently concerned about and barks at, then this is good information to help him become more comfortable with that stimulus in the future. This can be done by desensitizing him to the stimuli he is concerned with. Let us know if this is the case so we can help you desensitize him. If you are having a challenge getting him to walk on a leash, please let us know that too, and we will be happy to help you teach him how to walk on a leash nicely.

If you are carrying your puppy around, put him on the ground and let him start to experience different situations and life on his own. All four paws on the ground will help build his confidence. Confident dogs are not excessive barkers. They give alert barks to their family to let them know something is different or has changed. All they want is acknowledgment from you that they have done their job well.

Excessive Alertness

Some puppies take their protection role in the family a bit too seriously. Once your puppy has barked, alerting you to something, first check out what your puppy is alerting you to. There may actually be something you need to pay attention to.

Sometimes, just checking out what your puppy is barking at, saying words such as "Thank you," or "That will do," or "Okay" is all that is needed for him to know he has done his job. When you do give him a cue, say the cue in a soft voice. If you yell or speak too loud, he will think you are joining him in the bark and you too are concerned. Once you let him know in a quiet voice that he has done his job, simply walk away.

When you walk away and stop paying attention to him, there is a good chance he will stop barking. This is because you are no longer giving him attention. Your lack of concern about his reason for barking may serve as an example of "Oh, that must not be important," and the barking will stop.

If that is not enough, check out what he is barking at, and then call him over to you in a happy voice. At first, stand only a foot or so away from him. The second he turns to come to you, mark the behavior and encourage him to come to you with a healthy treat as the reward. You will soon find that when your puppy barks at something and you acknowledge his alert, he will simply stop barking.

Other Options

You can also interrupt your puppy's barking pattern by making a short, sharp, and/or unusual sound to distract him. A good example of this sound could be a plastic bottle with some loose change or rocks in it. Give it a single hard shake when your puppy is barking. The second he stops barking, give him a verbal cue, either "Thank you," "That will do," "Quiet," or "Okay." He will stop to see what the noise was. The second he turns his attention to you, by looking at you, mark the behavior with a word such as "Yes," or give a click from your clicker. Then reward him with praise, a pet, or a healthy food treat for looking at you and stopping the excessive barking.

You can also leave the room when the puppy is barking. By leaving the room, you are taking all attention away from him for his inappropriate barking. If he is barking at something outside, you can consider covering the windows where the behavior is occurring.

Head Collars

Having your puppy fitted for a head collar and then using the collar to address the barking behavior is another way to help him understand that you want him to stop barking once an alert has been given and acknowledged by you. Our office can fit your puppy with a head collar, but please make sure you check the fit frequently as puppies grow quickly and we do not want his collar to become too tight.

Once the puppy is comfortable with the head collar, attach your leash to the collar and stay with him. Create a situation or have a friend create a situation that would normally cause him to start barking excessively. Let him bark a few times, then give him the cue. You may use "Thank you," "That will do," "Quiet," or "Okay," or any other word(s) you would like, to consistently let him know he has done his job. Once you say the cue, gently pull the leash to the side to close his mouth. The second he stops barking, mark the quiet behavior with a "Yes" or a click from the clicker, and reward him with an ear scratch, a treat, or a pet while releasing the tension on the leash the second he stops barking. If he starts to bark again, repeat the cue and pull on the leash again until he stops barking. You will have to repeat this exercise several times a day over a few weeks until he understands that when you give the cue "Thank you," "That will do," "Quiet," or "Okay," he is to stop barking.

Biting and Nipping

Biting is a normal behavior for puppies. This is one way puppies play with each other to organize the hierarchy of the social group. So when you and your puppy begin your relationship, your puppy must be taught that your skin is much more sensitive than his. Play biting you is not an acceptable behavior.

Old techniques to stop this behavior, such grabbing your puppy's muzzle, giving your puppy a shake and saying "No," or pinning your puppy down, can only make matters worse! These types of reprimands can be interpreted by your puppy as an act of aggression. In many cases, reprimands such as these can even escalate the problem. Correction should not be handled with any methods the puppy could misinterpret. When hurtful methods are used to stop this behavior, play biting can quickly escalate to aggressive biting.

You will want to send a clear message to your puppy that biting you or other humans is not an acceptable behavior. You can accomplish this by withdrawing all attention from him when he bites you. Attention, whether positive or negative, is still attention. With that in mind, even scolding the puppy is a reward because he got your attention. When he starts to bite, walk away.

If your puppy is biting your shoe or pant leg, you may find it a little harder to walk away from him. When he grabs your pants or shoes, take him into a small area, such as your bathroom or another small room in the house. Re-create the behavior that caused him to bite you. When he begins to bite, do not say a word—just walk out of the room and close the door quickly.

Be careful not to close the door on your puppy's nose or paws; you will not want to hurt him when doing this. Leave him isolated in the room for a few seconds. Go back into the room and act as if nothing ever happened. Begin to re-create the behavior that caused him to bite again. When he tries to bite at your shoes or clothes, quickly walk out of the room again and close the door. After repeating this a few times a day over several days, the puppy will realize that every time he grabs your clothing or shoes or bites you, he will be left all alone. Through isolation and being ignored, the inappropriate biting behavior will stop.

It can sometimes get confusing to you and your puppy when playing together. Was the bite on purpose or was it an accident? To be safe, always assume it was on purpose. Stop the play and walk away from your puppy. Most bites are on purpose, so continuing the play will only confuse your puppy.

Try not to offer reasons for the biting behavior. Words such as "My puppy did not mean it," "It was an accident," and "It was really my fault, not my puppy's" can cause more harm than good. Trying to be understanding and helpful by offering reasons why the behavior happened can only make matters worse and confuse your puppy.

Your puppy may think biting is a wonderful game. If that happens, the biting behavior will increase in intensity as you continue to allow and reinforce it. The more demanding your puppy is with this wonderful game, the harder he may begin to bite. Before you know it, the biting will escalate, and the puppy will think it is part of the game and biting you is an acceptable behavior.

Another thing you can do is put your puppy in the crate for a time-out. Wait for him to settle down and then open the crate slowly to let him out. If he tries to bite at you again, simply close the crate door and leave him in the crate for a little longer. Then wait for him to settle down and try to open the crate again. You may have to repeat this exercise many times a day over a few days for your puppy to understand.

When putting your puppy in the crate or into his space for a time-out, do not drag him by the collar. When he is dragged by the collar, he may become afraid of your hands reaching for him. This technique could result in your puppy being afraid of hands and cause him to shy away from you when you try to reach out to pet him. Instead, either pick your puppy up, if you can do so without getting hurt, or have a short, two-foot leash attached to his collar. This way, when he bites you, you can use the leash to bring him to his crate and lead him.

Remember, do not open the crate for a barking dog. When you do, you are teaching your puppy that you can be trained. Many puppies are capable of barking for a very long time. This is how they wear down their owners. But you must use tough love here if you want him to learn quickly. When the puppy carries on for 30 minutes, it is tempting to let him out for a little quiet in the house, but do not let him out. If you do, the puppy will only bark longer next time and matters will get worse, not better. Always wait for that moment of silence before opening the crate for a barking dog.

Hugging, kissing, and holding your puppy is a human need, not a puppy's. Many puppies do not like to be held quite as long or as often as many puppy parents might think they do. You should not expect acceptance of hugs and kisses from your puppy. Your puppy needs time to understand you and your ways. Many puppies will allow the hugs and kisses once they are comfortable with their new family.

Since he is constantly learning, he will figure out that kissing, holding, and hugging will not harm him, especially if you listen to his request to stop or be let down when he asks. By giving him the time he needs to get used to some of our human idiosyncrasies, he will become comfortable with this human hugging and kissing thing. After all, how comfortable would you be if someone walked up to you and smelled your butt? That is a natural behavior for dogs, just like hugging and kissing is for us.

Puppies can interpret being held as being confined, which is something they do not like. To ask you to let them down, some puppies will start off licking your arm or hand—that is their way of politely asking to be let down. When the request is missed by you, the only option your puppy has to let you know he wants to get down is to start biting at your hand, arms, or clothing. By paying attention to your puppy's signals, you can avoid this situation.

If, after trying these methods, your puppy is still biting, please call our office. There may be a physical or emotional problem going on that should be addressed immediately.

Bolting Out Doors

Bolting out the door is a very dangerous behavior for both your puppy and the people he may meet. He can get lost or hurt, or if someone approaches him, he may bite out of fear.

The step-by-step process outlined below to address this behavior is based upon the *sit* and *stay* cues. Remember that it is important to keep eye contact with the puppy when training *stay*.

Place a small rug near the door your puppy has tried to bolt out of. Make sure the rug is far enough away from the door so you can open and close the door without the puppy having to move. Practice *sit* and *stay* on this rug with your puppy. Remember to always release the puppy from a *stay*. If your puppy bolts out of every door in your home, then start off with one door until he can hold his *sit/stay*. Once he is doing well at one door, you can train the same behavior again at another door in your home. Usually after two or three doors, your puppy will understand that "sit/stay" means "sit/stay no matter which door he is at."

Once he is doing well with the previous step, you can introduce the door actually being opened. Put him on his leash and ask him to "sit/stay" on his rug by the door. Hold onto the leash or step on the leash, open the door, and immediately close the door. Return to the puppy, mark and reward him if he held the *sit/stay*, and release him. If he did not hold the *stay*, put him back in the exact same position he was in originally and give the cue for "sit" and then "stay." Wait a moment, then mark and release him for staying still. Repeat this exercise until you can open and close the door quickly without him breaking the *sit/stay* cue.

Gradually increase the amount of time you can hold the door open without him breaking the *sit/stay*. Each time, return to the puppy, mark and reward him if he held the *sit/stay* position, and release him. If he did not hold the *stay*, put him back in the original location and into a *sit/stay* again. Wait a moment, then mark and release him. Repeat this exercise until you can keep any of your doors open for 30 seconds and your puppy holds the *sit/stay* cue. Remember to always mark, reward, and release him from the cue.

Now have a friend ring the doorbell or knock on the door. Your puppy will probably run to the door to see who is there. Calmly walk to the door with leash in hand, call your puppy's name to get his attention, and put his leash on him. Put him into a *sit/stay* on his rug, hold onto the leash, and open the door slowly. If he holds the *stay*, have your friend come in to the house, return to the puppy, mark and reward him with a jackpot, then release him. If he did not hold the *stay*, put him back in the original location and into a *sit/stay*. Wait a moment, than mark and release him. Repeat this step until the puppy can hold his *sit/stay* until your friend is inside of your home and then quickly release him. At this stage, ask your friend to ignore the puppy completely and not make eye contact. You want your puppy to keep eye contact with you until he is released.

Then repeat what you did in the last step, but ask your friend to calmly greet your puppy while he is in the *sit/stay* position. If he breaks the *sit/stay*, your friend should immediately ignore the puppy while you put him back in the original position and into a *sit/stay*. Have your friend try to calmly greet your puppy again, and this time offer him treats while your friend goes to pet him. If he can hold his *sit/stay* while being petted, quickly mark and reward him.

Repeat this a few times until he is consistent at paying attention to you while your friend pets him. Once he is consistent, you can stop offering treats while he is in the *sit/stay* and have your friend softly pet him. If he holds the position, jackpot him with many small treats for a job well done. If he is still breaking the *sit/stay*, ask for a *sit* and then quickly reward and release him. For now, training the *sit/stay* at the door is done. By ending this training session on a positive note, your puppy will have an opportunity to think about what just happened and look forward to the next training session. You can work on this behavior later in the day or the following day. Keep your lessons short, and always end a training session on a positive note.

Chewing

Chewing is a natural behavior for puppies and adult dogs. It will be important to have safe toys for your dog to chew on throughout his life. Puppies will put just about anything in their mouths, so it is your job to make sure your puppy knows what is appropriate to chew on and what is not.

With a new puppy in the house, it is a good idea to puppy-proof your home. This can protect your puppy from hurting himself and save your shoes, clothing, pillows, toys, and your home from destructive chewing.

Plush toys and most squeaky toys are not much help in relieving the discomfort of cutting new teeth, nor will they satisfy the puppy's need to chew. If you do not want the puppy to chew on shoes, do not give him an old shoe to chew on. The same holds true for all other items in your home. The puppy cannot distinguish between old or new, and expensive or inexpensive, so make sure chew toys are chew toys and not other items.

If your puppy has something in his mouth that should not be there, offering him a safe toy in exchange for the inappropriate item can be your first step in addressing this behavior. However, you cannot spend the rest of your life walking around the house trading items for toys with your puppy. You will want to teach your puppy what the cue "Leave it" means. You can teach him what this cue means in a few different ways.

For the first method, you can show the puppy a treat and hold it in your hand or put it on a chair or coffee table. When the puppy goes toward the treat, say "Leave-it" in an authoritative voice. If your puppy tries to get the treat, move the treat out of his reach quickly. If your puppy stops the forward movement, and turns his attention toward you instead of the treat, mark and reward him with a different treat and remove the treat you asked him to leave alone.

Repeat this exercise several times over many days until you can put the treat right on the coffee table or right in front of the puppy's nose and say, "Leave-it," and he looks at you instead of the treat. When he leaves it completely alone and does not try to take it, and he looks at you consistently, he is now beginning to understand what the cue "Leave it" means.

Once your puppy is good about leaving the treat alone while you are right next to him, try placing the treat on a table and taking one or two steps away from the treat. When your puppy goes near the treat, give the cue "Leave-it," but stay close to the treat in case he tries to grab it. If he goes for the treat, say the cue "Leave-it" quickly in an authoritative voice. If he stops going for the treat, mark the behavior with "Yes" or a click on the clicker and reward him with a different treat. If he does not stop going for the treat, quickly remove the treat before he gets it. Repeat this many times until you can give the cue "Leave-it" and take a few steps away without him trying to grab the treat. In time, he will learn that when you give the cue "Leave-it," he is to drop whatever he has in his mouth or simply not put the object in his mouth in the first place.

A second way you can teach your puppy the cue *leave-it* involves a soft, rubbery treat (like a piece of cooked chicken hot dog). Show the puppy the treat and then place the treat under your shoe. Make sure you are wearing close-toed shoes. At first, your puppy may scratch and dig at your shoe with the treat under it to try and get it. Do not say a word; stand up straight and wait. The second he looks at you instead of the treat, mark and reward the behavior with a different treat. Repeat this exercise many times until he consistently looks at you instead of trying to get the treat from under your foot. Never give the puppy the treat you want him to leave alone, or you will confuse him. You can use the treat under your foot later on, but not while you are training this behavior.

Once your puppy stops trying to get the treat from under your shoe and looks up at you consistently, you can take this a step further. Next time, put the treat on the floor under your shoe while saying "Leave-it" in an authoritative voice. Uncover the treat so your puppy can see it. If the puppy tries to get the treat, cover it with your foot immediately so he cannot get it. Wait a few seconds and try again. Repeat this exercise until your puppy looks at the treat on the floor, ignores it, and looks at you instead.

Once your puppy has been successful with ignoring the treat, it is time to make it a little more challenging for him. This time, throw a few treats on the floor and say "Leave-it." Make sure the treats are close enough together so you can cover them quickly if he tries to get one of them. Repeat this exercise several times until your puppy can look at all those wonderful treats on the floor, ignore them, and look at you. Once he has successfully accomplished this many times over many days, he will know what to do when you give the cue "Leave-it." "Leave-it" means, if it is in his mouth, he is to drop it. If he is thinking about putting something in his mouth, the cue "Leave-it" also means do not to touch that item.

The steps in the second method above are nicely demonstrated in the Puppy Smarts Chewing video.

Dogs enjoy chewing throughout their lives. It is important that your puppy and adult dog have safe toys and bones to chew on. Hard bones, rubber toys, and toys that you can put some treats in for the puppy to work for are perfect chew toys. Besides meeting his need to chew, toys you can place treats in will also offer him mental stimulation.

If your puppy or adult dog seems to be chewing excessively, there may be other reasons for this behavior. Your puppy may have a dental problem that needs to be addressed with your veterinarian. If he is between six and nine months of age, his permanent teeth are erupting, and chewing may increase dramatically. During this stage in your puppy's life, it will be a good idea to confine him to his crate when he cannot be supervised, even if he has been trustworthy in the past. This is an age when chewing is at the top of his priority list, and just about anything can be fair game.

Collars

Many puppies do not like their collars, so it is a good idea to give them time to get used to the smell and taste of the collar first. Hold the collar in your hand and let your puppy sniff and put it in his mouth if he wishes to check it out. When he begins to get bored with the collar, take it away.

Next, let him see the collar again and have some wonderful, tiny, tasty treats while you simply drape the collar over his neck. If he is apprehensive, stop. Wait a few minutes and try again. Drape the collar over his neck while distracting him with a tasty treat. Move slowly and with confidence. If he leaves the collar draped over his neck for a few seconds, mark and reward his bravery by telling him what a brave little boy he is and giving him many tiny treats. Repeat the exercise several times until he becomes comfortable with the collar being draped over his neck.

The next time you work with the puppy on wearing his collar, show him the collar and offer him a tiny treat. Then place the collar between your two hands and go under his neck to give him a nice neck rub for just a few seconds. Repeat the exercise, and this time clip the collar on his neck. If the collar has a clip that will make a noise, try to cover the sound by catching the clip before it opens and closes in place. If it is a buckle collar, put one end through the other and quickly fasten it. If the puppy pulls back, stop. Wait a few minutes and give him another nice little neck massage while holding the collar between your two hands for just a few seconds. Tell him what a great little boy he is and walk away. For now, that is enough training.

Later on, try again. Start with the collar between your hands and with confidence place it around his neck and close it. If he struggles, stop what you are doing and walk away from him. Come back in a few minutes and try again. This time, offer him a treat and place a few more treats on the floor for him to eat. While he is eating the treats, place his collar on his neck for just a second. When it slides off, pick it up and hold the collar between your two hands. Drop a few tasty treats on the floor, rub his neck, and attach the collar. If he is still concerned, stop and go back to draping the collar over his neck for a second, giving him a treat, and telling him what a good boy he is. For now, the lesson is over.

Over the next couple of days, repeat the exercises above, starting from the last place he felt comfortable with the collar touching him. In time, he will let you put the collar on.

Simply demanding that the collar goes on now works much faster, but you will have to make the decision whether you want to build this relationship on trust or fear. When you stop doing something your puppy is afraid of, it is a gentle way to let him know you understand and are listening to him. When you listen to your animal's concerns, you are building a stronger relationship that is built on trust.

Come

Come can be an easy cue to teach your puppy and should be turned into a game to make it fun. Teaching your puppy to *come* on cue will protect him from danger throughout his life. There are two important rules to consider when training the cue *come*.

- Never scold your puppy when he has come to you, no matter how slowly he comes.
- Never call your puppy by his name when you will be doing something he may not enjoy, such as giving him medication or giving him a bath.

There are a couple of different options for training the cue *come*. In the first option, when your puppy is lying down quietly, show him a tasty treat in your hand (lure) and walk about three feet away from him. In a happy voice, say his name and the word "Come" (cue). The second he takes his first step toward you, quickly mark the forward movement with a word like "Yes" or a click from your clicker. Encourage him to come over to you to get the tasty lure. When he does, quickly mark his forward movement and reward him with the treat when he reaches you. Give him some special attention by playing with and rubbing him, and tell him he was a good boy in a pleasant and happy voice.

Later on in the day, when he is lying down comfortably, repeat the exercise. Show him the treat, say his name in a happy voice, and give the verbal cue "Come." Remember to mark his forward movement immediately, and give him the treat when he gets to you. Repeat this exercise several times a day for the next few days.

Once your puppy is coming to you consistently on cue, you will want to lose the lure (showing him the treat), and replace it with the reward (not showing him the treat). (For more information about the difference between lures and rewards, ask your Patient Behavior Advocate for the Lures and Rewards handout.) To accomplish this cue without a lure, begin with your hands in the same position as they were when you showed him the treat, but do not have a treat in your hand. Use his name first then give him the cue "Come" in a happy voice. Mark the immediate forward movement with a word or a click, and give him a jackpot of treats as his reward when he reaches you. After he has successfully *come* on cue three or four times, you can begin to extend the distance between the two of you.

As part of the second option, if you have more than one person in your home, you can play monkey (puppy)-in-the-middle. Have two people stand a few feet apart and take turns calling him. Both people should have many small treats to offer him as the reward for coming. For the first two or three times playing this game, use the reward as a lure until the puppy gets the hang of the game and comes quickly to the person who calls him. The person who calls the puppy should use a happy voice when saying his name along with the cue "Come." The person who is not calling him should stand up straight and ignore him.

Once he has successfully come to both people a few times in a row, extend the distance between the two people a few steps at a time. In the beginning, it is okay to show him the lure treat, but as time goes on, it is important to turn the lure into the reward and hide the treat. In the future, you do not want him visibly checking out your hands from a distance to see if you have a treat before he is willing to come to you. Before you know it, he will be flying back and forth between the two of you and having a wonderful time as he is learning the new cue.

You will want to repeat this exercise many times with different family members until he comes to all family members consistently. Once he does, he is ready for the next step in training this behavior.

The Next Step

In the next step, it is time for a game of hide-and-seek. While your puppy is in one room of the house and you are in another, say his name and give the cue "Come" in a happy voice. Use your happy voice to encourage him to come find you. The second you see him, mark his forward movement and give him a jackpot of many small treats when he gets to you. This was much more difficult than when he could see you, so let him know how proud you are of him.

Repeat this exercise a few times a day over many days. When he comes to you quickly every time you call his name, it is time to start adding distractions.

Adding Distractions

A distraction can be a toy, another person, or anything your puppy will want to pay attention to more than to you.

Give your puppy a toy to play with and walk a few steps away. Say his name and give the cue "Come" in a happy voice. When he looks up at you, encourage him to come to you. Mark any forward movement, and offer him a jackpot of rewards. Then, tell him to go play, or use another word that releases him. This is to let him know that even though you may be interrupting his playtime, he can get a treat from you and go back to playing with his toy. Repeat this a few times a day over the next 10 days, extending the distance between the two of you until you can give him a toy, walk out of the room, call his name with the cue "Come," and he drops his toy to see what you want or have for him. Once he is consistent at coming to you when he is playing with his toy, you can now add another distraction—a person.

Ask someone to start playing with the puppy. Then, standing just a few feet away, say his name in a happy voice and ask him to *come*. If he looks at you, encourage him to come until he reaches you, and offer him a jackpot of treats and lots of petting, and tell him what a good boy he is.

If he is too distracted with the other person, walk over to him and lure him. Let him smell the treat for just a second and walk away. Encourage the other person to continue playing with him. Then, quickly call his name and offer the cue "Come" in a happy voice. Do this quickly, as you do not want him to forget you have that wonderful treat waiting for him. Once he is consistently coming to you while someone else is playing with him, begin to extend the distance between the two of you until you can be at the other end of the house, call him by name and give the cue "Come," and he will come even though he is getting attention from someone else. You will want to repeat this exercise many times with different family members until he comes to all family members consistently.

The next step in training this behavior is with other animals in the house. You will take the same steps as you did with the other distractions. Always move at your puppy's pace—do not progress too quickly. You want him to be successful when learning new behaviors, and failure can result if the training process is rushed.

Distractions Outside the House

Once your puppy is consistent with handling distractions inside, it is time to move the training outside. When you are beginning to train him to *come* on cue outside, you will always want to keep him safe.

Dogs do not generalize well, which means the *come* cue may be brand new to him outside. With a loose leash on the dog, face him and take a few steps backward. Say his name and give the cue "Come" in a happy voice. The second he looks at you, encourage him to come; then mark and reward him for coming. Once he does come to you, it is important to release him to go play again. Words such as "Go play," or "Free dog," will work as a release—just keep the release consistent throughout the puppy's training.

Repeat the "Come" cue a few more times outside until your puppy comes to you consistently. As he becomes more consistent, you can use a long line to extend the distance between the two of you. At this point, the behavior is not as established as it should be for him to be allowed off-leash. Taking your puppy off-leash too quickly can lead to many problems and could be dangerous for him. Being outside without a leash or long line is a privilege that must be earned. If you take him off-leash before the puppy is ready, he may run away from you to play. Once he figures out he can run away from you when he does not have a leash on, training the *come* cue and other behaviors becomes more difficult. Wait to unleash him until he is consistently coming to you outside with many different distractions.

After a few months of consistent behavior, you can begin training him to *come* on cue off-leash in a fenced area. When you take him off-leash, ask for the *come* while close to him and slowly extend the distance a foot or two at a time.

A word of caution: Avoid training the *come* behavior too many times in the same day or having the training sessions occur too close together in time. You do not want to habituate him to the cue *come*.

It is very important that the puppy is always rewarded for coming to you, no matter how long it takes. Coming to you should always be a pleasant experience so that in time he will know coming to you is always an opportunity for a nice treat or that something good will happen. Once he is conditioned to *come* on cue, you can begin to offer the rewards intermittently.

Crate Soiling

Crate soiling can be a real challenge for many new puppy parents. Know you are not alone and we are here to help you. Crate soiling can happen for many reasons. For example, your puppy may have been forced to live in the crate full-time prior to arriving at your home. In this situation, he has never had a choice on where to eliminate. As a result, he now may have a preference to eliminate in his crate. Among other reasons for crate soiling are the following (ask to see the Crate Training handout for more details):

- *Your puppy's crate may be too large.* This can be easily addressed by getting a smaller crate that allows your puppy only enough room to stand up, turn around, and lie down comfortably. Alternatively, you may reduce the size of your large crate by placing a space-occupying object, such as a piece of wood, cardboard, or other material, in the crate. There are crates on the market today that have a wire wall that can be adjusted to fit your puppy's size. As your puppy grows, you must move the wire barrier frequently to accommodate his increasing size. Another option you can choose is to create an elimination area in the larger crate. This can be accomplished by placing a wee-wee pad in a doggie litter box. Place the litter box at one end of the crate and give him sleeping quarters at the other end of the crate. This will allow your puppy to eliminate in one area and sleep in the other, making cleanup much easier.
- *You forgot to let your puppy eliminate before placing him in the crate.*
- *You left the puppy in the crate too long.* Puppies can wait for only short periods of time before needing to eliminate. If your work schedule prevents you from letting your puppy out of the crate sooner, consider a pet sitter or doggie day care.

Once you have adjusted the puppy's crate for his size, tried getting a pet sitter to let him out in the middle of the day, and even tried doggie day care, then here is your last step for working on this problem. Keep him in his crate while watching him. Every time the puppy seems to be sniffing the floor in his crate, looking for a place to go, or circling in his crate, and he has been in his crate for more than three or four hours, he may need to relieve himself. Take him outside to the designated area, stand there, and wait. The minute he eliminates, mark and reward him right there with a very special treat he only gets when he relieves himself outside or in the designated area inside. Over time, he will become conditioned to want to go outside so he can get one of those special treats.

This can be a challenging behavior to deal with, but with time, patience, and training consistency, your puppy will learn not to soil in his crate.

Crate Training

Crate training is an extremely valuable tool for you and your puppy. You will reap great rewards throughout his life by training your puppy to be comfortable in his crate. The crate will become his bedroom. It is a haven where he can get away from energetic children and company. It is a place where he can rest and be left alone. Crate training is the easiest way to control your puppy's environment, and it is helpful in housetraining.

Placing your puppy in his crate is not the same as leaving him in a laundry room or kitchen. Those areas are used and shared by the family, and your puppy needs and deserves his own space.

Children should be told to leave the puppy alone when he is in his crate. More important, children should never be allowed to go into the puppy's crate.

Your puppy's age can be used as a general rule to determine how long the puppy can stay in his crate before needing to relieve himself. This information holds true while the puppy is awake or active. Using your puppy's age in months and adding 1 will give you the number of hours he can be kept in the crate before needing to relieve himself. For example, a two-month-old puppy should not be left in his crate for more than three hours, and a three-month-old puppy should not be left in his crate for more than four hours. This rule holds until the puppy is about six months old, when the puppy can be left in his crate for six to eight hours before he will need to relieve himself. It is not appropriate for any dog to be left in a crate for more than 10 hours a day. Your puppy needs and deserves exercise time, playtime, socialization time, training time, and the opportunity to interact with his new family. A crate is a training tool and should be used as such until the puppy understands all the rules in his new home. As he gets older, unsupervised time alone outside of his crate should be increased gradually.

A very important thing to remember is to never let your puppy out of the crate when he is barking or crying. If you do, you have allowed the puppy to train you! If the puppy has been in his crate for a while and starts to bark, he may need to go outside to relieve himself. Wait a few seconds until he stops barking or whining and then quickly open the door and take him to the designated elimination area. Immediately mark and reward him for eliminating in the proper location.

Begin crate training by setting up the crate in a room where the family is usually present. Place a dog bed or soft blanket inside the crate for the puppy. Leave the door open for a few hours and give him time to get comfortable with the look and smell of the crate.

Once he appears to lose interest in the crate, throw a few tasty treats inside the entrance of the crate to lure him into it. Tie the crate door securely open to ensure the door does not close accidentally and frighten him in the process of learning to be comfortable with his new crate.

If your puppy has not eaten the treats within 15 minutes or so, pick them up. At your next scheduled feeding, place your puppy's food bowl inside the entrance of the crate. Walk away from the crate and watch him.

If your puppy is hungry, he should approach the crate to eat. If after 15 minutes he still refuses to go near it, take the food out of the crate and place it two or three inches outside of the crate door.

Again, wait about 15 minutes to see if he will eat the food. If he will not go near it, move the bowl a foot or so away from the crate and repeat the exercise. Keep doing this until you find a place where he is comfortable enough to eat.

Once he is comfortable eating where the bowl or treats are placed, gradually move the bowl or treats closer to the crate and eventually into the crate and toward the back. Always leave the door tied open so the puppy can go in and out of the crate by himself during this phase.

Next, place a toy with treats inside it in the crate. This time when your puppy goes into the crate, close the door for just a few seconds. Then open the door and let your puppy out of the crate.

Repeat this exercise a few times a day, gradually increasing the length of time you keep the puppy in the crate with the door closed. Once he is comfortable being in the crate with the door closed for about 30 minutes, let him out of the crate. Remember to not open the door of the crate for your puppy when he is barking or crying. If you do, the puppy will learn that if he makes noise, he will be let out of the crate. This is not the lesson you want him to learn.

Most puppies love their crates. There are, however, some puppies who are very afraid of their crate and want nothing to do with it. Under these circumstances, it is better to find another method of confining a puppy than it is to force him into a crate. Other confining options may include a utility room, a bathroom, an exercise pen, or a baby gate in the doorway of a small room. For small-breed dogs, perhaps a baby's playpen would do.

Many dogs from shelters and pet stores have been confined to crates for extended periods of time. These puppies may associate a crate with a negative experience and are often very concerned about being placed in a crate again.

Crate training has many rewards, including being able to leave your puppy alone without any damage to your belongings or accidents on the floor from improper elimination. This will speed up the housetraining process and provide your puppy with his own secure and comfortable bedroom while managing his environment.

Down

There are three basic steps when teaching your puppy this cue:

- *Lure.* Put the puppy into a *sit* position and then place a treat in your hand in front of the puppy's nose. Move the treat slowly down between his front legs to the floor. When most puppies' noses go between their front legs, the back legs slide back and they go down, which is exactly what you want. A slippery surface works best for this training, as it offers little resistance to the puppy sliding down to the floor.

- *Mark.* The second the puppy's body hits the floor, say a word such as "Yes" or give a click from a clicker to mark the behavior. This lets him know that was the behavior you wanted.
- *Reward.* Give the puppy a different tasty treat in addition to the one you just used to get him into the *down* position. Make sure he receives the treat while his body is still on the floor, or you will be rewarding the wrong behavior. The correct position is body on floor to receive his reward.

Once the puppy is consistent at downing promptly every time you show him the lure, change the lure into a reward. This can be accomplished by moving your hands the same way, but without any treats. Add the verbal cue "Down" at the same time you are moving your hand to the floor. The second his body hits the floor, mark and reward him quickly. With a little time and practice you can stop using your hands entirely and merely give the cue "Down." The finished behavior has three basic steps:

- *Cue.* Say the cue "Down."
- *Mark.* Say "Yes" or a click the moment the puppy's body hits the floor.
- *Reward.* Give the puppy a treat.

It is very important to change the lure into a reward only for completing the behavior, or the puppy will listen to you only when he sees a treat. The first time he *downs* without you showing him a treat, jackpot him for a good job. A jackpot is many tiny treats given to him in a row. Instead of just one treat, a jackpot will be four or five tiny treats in a row. This is to let him know he did a great job.

Helping Your Puppy Generalize the Cue

If you are teaching your puppy to *down* in front of you, start asking for the *down* at your side. This will be a new behavior you are requesting since dogs do not generalize well. When you teach your puppy to *down* at your side, train the same way you did when asking for the *down* when he was in front of you. Once he is consistent in front of you and at your side, it is time to introduce new environments, such as a different room, then outside, then from across the room, and so on. Each step takes time. This is a progression toward teaching your puppy that when you give the cue "Down," he learns in many different places and from many different positions that "Down" means you want his body on the floor.

Often, a new puppy parent thinks its puppy is stubborn, hardheaded, or has selective hearing because the puppy will not *down* when asked for the cue. In many cases, it is because the puppy parent did not teach the puppy the cue in many places or with distractions. Another reason why some puppies do not *down* on cue is because they were habituated to the cue. This can happen when puppy parents repeated the cue more than once by saying "Down, down, down." This can confuse the puppy and he will get used to hearing "Down" (one time) with no behavior required of him, which makes the cue ineffective. If you must repeat the cue, take a step to the right or left first, get your puppy's attention, and then repeat the cue—once.

Troubleshooting *Down*

Some puppies resist the above method. You can address this resistance in a couple of ways.

The first option is to begin by sitting on the floor with your puppy between your legs. Bend one leg up and slide your hand under your bent knee with the treat in it. Show your puppy the treat. As your puppy puts his head down toward the treat, bring it under your leg so the only way he can get the treat is to lie down and get under your bent knee. This may be a little difficult, but you should have to use this method only a few times before he is willing to *down* on cue.

Once your puppy is consistently downing from this position, you can then train the behavior with him next to you on the floor instead of between your legs. From that point, ask for the behavior while you are kneeling on the floor and then from a standing position. Remember to always mark and reward when he gives you the desired behavior.

Another way to train this behavior is to pay attention to your puppy. When he goes to lie down on his own, say the cue "Down," mark the behavior, and give him a treat to reinforce the *down*. When you repeat this several times over several days, the puppy will learn the cue *down* on his own with a little help from you. Remember, once he is downing consistently, move to different rooms and add distractions while perfecting this behavior. This method of training is about catching your puppy doing what you want, giving it a name (cue), then marking and rewarding to establish the new behavior.

If you find that your puppy seems to be resisting the *down* cue, do not push his body down to the floor. There may be a medical reason for this resistance. Instead, contact our office and set up an appointment. Many puppies that have resisted the *down* cue were later diagnosed with a physical problem.

Drop-It

Drop-it is a great cue and should be taught to every puppy. Trying to pull items out of a puppy's mouth can be challenging, dangerous, and misunderstood by the puppy as an act of aggression by you. Teaching *drop-it* starts off as a game of exchange: I will give you this if you will give me that.

When you begin to train the cue *drop-it*, make sure you have some wonderful treats to give the puppy in exchange for what he has in his mouth. When he is playing with a toy, walk over to him and give the cue "Drop-it," and offer him a piece of chicken or other wonderful treat in exchange for his toy. The second he drops his toy for the treat you have offered, mark the behavior with "Yes" or a click from your clicker and give him the treat. This will teach your puppy two things at the same time: The puppy will learn to drop objects in his mouth when you give the cue "Drop-it" and it will also help him not to guard his toys.

Repeat this many times with many different safe puppy toys over the next couple of weeks. If he grabs other items in the house, use these items to train this behavior as well. If he releases the item in his mouth when he hears the cue "Drop-it," mark the behavior and reward him by giving him one of his toys in exchange for the item he should not have in his mouth. Do not begin training with objects you are having trouble getting him to release. Start off with easy objects and work your way up to the more challenging ones. You always want to set him up for success and then build on those successes. That is what is called shaping a behavior. You start off with small successes until he is doing exactly what you request.

As your puppy becomes faster at releasing what is already in his mouth, begin to offer the exchange less frequently. For example, the next time you say "Drop-it," he may get an ear scratch and be told he is a good boy. The next time, he might get a treat, and the next time just a verbal "Good boy." Yet another time he might get three treats instead of just one. By keeping him guessing on whether he gets a treat, an ear scratch, or kind words, you will keep his attention on you and make it enjoyable to listen to the cues you give him. Once you feel he really understands what the cue "Drop-it" means, you can start adding distractions. This could be other people in the room while you are training him, or perhaps outside in a safe, enclosed area.

After a while, the puppy will understand what "Drop-it" means. This will make playing fetch with him easier to train since he will already know the cue.

If, for safety's sake, you must physically take something out of your puppy's mouth, gently take your hand over the puppy's muzzle and, using your thumb and index or middle finger, gently squeeze at the back of the jaw between the upper and lower teeth. Reach inside his mouth and take the object out of it. Once the object is out, tell the puppy what a good boy he was and give him a treat. This procedure should never be done in a mean or hurtful way and should be used only when the puppy's or another's safety is at risk.

If your puppy is growling at you, an alternative way to get something out of his mouth is to toss a tasty treat a couple of feet away from him. When he drops the item to go get the treat, pick the object up quickly and put it away or throw it away.

Exercise and Play

Puppies need both mental and physical exercise. When you meet these needs, your puppy will be easier to manage. If these needs are not met, your puppy can become very bored and create a lot of mischief. This can cause problems for both you and your puppy.

Taking the puppy for a walk a couple of times a day is a good beginning, but this is not enough. If it is possible, find a place where he can be taken off his leash safely to let him run around and play. He will need interaction with you to really get the exercise he needs. This is a great time to introduce him to a few games that will stimulate him both mentally and physically.

When you take him to a safe place where he can be off-leash, bring one of his favorite toys along to play fetch. This could be a squeaky toy, a tennis ball, or anything else he enjoys playing with. Bring plenty of his favorite treats as well.

There are many wonderful games you can play with your puppy to stimulate him mentally and also give him the exercise he needs. If he has a safe, properly immunized puppy friend or adult dog to play with, this can be a great way for him to exercise. He will also learn how to play nicely with other dogs. Playing games are a fun way your puppy can learn new cues and exercise at the same time.

Fetch!

Let the puppy see the toy right before you take him off-leash. In a happy voice, ask if he wants the toy while bouncing or squeaking it. Throw the toy just a foot away from where you are standing and say the cue "Go get it." The second he puts the toy in his mouth, mark the behavior with "Yes" or a click from your clicker. Then give the cue "Bring it here," and lure (let him see the treat) him over with a wonderful treat like a piece of chicken.

Now give the cue "Drop-it." When he does, give him the tasty piece of chicken and tell him what a good boy he is. Repeat this, but throw the toy just a little farther each time, using the same cues. After throwing the toy five or six times, end the game while you are both enjoying yourselves. The next time the puppy has an outing with you, you can play again. Start off with the toy close, and each time throw it a little farther away. It is important to end the play training on a positive note while you are still having fun. If your puppy bites your hand by accident before dropping the toy, the game is over—no excuses. Put him back on-leash and walk away. Wait at least five minutes before beginning the game again or paying any attention to him. This is very important. You do not want to teach him it is okay to bite you under any circumstances.

Chase!

Another game you can play with your puppy while he is off-leash in a safe place is a game of chase. There is, however, one very important rule to this game: *Never* chase him. If you do, you will be training your puppy to run away from you.

With his leash still on, give the cue "Come" in a happy voice and start to run backward a few feet away from your puppy. (Be careful not to trip or hurt yourself.) When he follows and catches up to you, stop and mark his coming toward you with "Yes" or a click from your clicker, and give him a treat. You can repeat this a few times, extending the length of space between where you start and where you stop. Remember to mark his coming toward you each time, and then give him a treat when he reaches you to reinforce the desired behavior.

Note these few rules when playing with your puppy:

- If he bites you by accident, the game stops immediately. No excuses.

- If he jumps on you because he is excited, turn your back on him or step into his space. Once all four of his feet are on the ground, you can then mark and reward him for coming. Never give him a treat unless all four feet are on the ground. If you do, you will be rewarding him for jumping on you.
- Never just pull a toy out of his mouth. Exchange toys for treats using the *drop-it* cue until he understands the cue.

Tug-of-War!

Tug-of-war is another game you can play with your puppy. If he already knows the cue *drop-it*, this is a great game to play. If he does not understand the cue *drop-it*, then this game can be a great way to teach him the cue. Some people believe that playing tug-of-war with their puppies will make them aggressive. This is untrue. This game simply requires rules that need to be followed.

Find a safe toy that is made for this game. A safe toy allows room for the puppy to grab the toy without grabbing your skin. Play the game with him for a few minutes and then give the cue "Drop-it" and offer him a treat as an exchange for the toy. If he does not drop the toy, the game is over. Walk away and ignore him. If he bites you by accident while playing this game, stop and walk away; the game is over.

There is no such thing as an accidental bite. If you make excuses for the puppy that the bite was an accident, he will learn to bite again next time you play the game. After all, biting each other is one way puppies play together. He will not understand he should not bite you unless you let him know that all play and attention stop when a bite occurs.

In a few minutes you can try again, but for now you want him to understand that when he bites, the game stops and you will completely ignore him. In time, he will understand your rules and begin to play this game politely.

Other Activities

You can enroll your puppy in classes for exercise and play. You can choose from Agility, Fly Ball, Earth Dog, Freestyle Dance, or Frisbee. If you have access to a pool your puppy can swim in, that is also a great way to exercise him. If you do allow your puppy in a pool, make sure he understands how to get out of the pool safely on his own right away. This is a safety "must do" when a puppy is allowed in a pool. All of the above-mentioned exercises are great ways to give your puppy the exercise and exposure he needs. Introducing him to new environments and situations are a plus. These all provide physical and mental stimulation.

Grabbing or Training *Leave-It*

Depending on what your puppy is grabbing, there are a few ways to address this behavior. Grabbing can include taking clothing, food from children, treats, pillows, plants, and toys, among other objects.

Leave-It

When your puppy is grabbing at almost anything he should not be, a good cue to teach is *leave-it*. This is nicely demonstrated in the PuppySmarts Chewing training video. Once the puppy understands what the cue *leave-it* means, you can begin using the cue whenever he goes to grab something he should not. Each time your puppy obeys the cue, mark and reward the appropriate behavior. It is best to train this cue while the puppy is playing nicely with one of his toys. Simply walk up to him and say "Leave-it," and offer him a nice treat in exchange for the toy. Once he releases the toy, give him the treat. When he is done with his treat, give him back his toy to play with. Repeat this a few times a day until every time he hears the words "Leave-it," he immediately drops the item he has in his mouth.

Clothes and Shoes

When your puppy is grabbing at clothing and shoes, there is another training method you can use to address this behavior. Put your puppy in a small room (like a bathroom) and re-create the behavior. For example, suppose the puppy grabs shoes or pajamas when you walk by him. In the small room, try to walk as you did the moment before he grabbed your shoes or clothing. The second he grabs the article, quickly walk out of the little room and close the door. The puppy will let go of the item when the door is almost closed. Be careful not to get the puppy's muzzle in the door jamb; you do not want to hurt him.

Once you close the door, walk away for 10 or 20 seconds. Go back into the room and act as if nothing happened. Talk and play with your puppy as you would normally do to create the grabbing behavior once again. Once the puppy grabs the article, walk out of the room quickly and close the door. Do not say a word. After repeating this exercise five or six times, he will figure out on his own that when he grabs your shoes or clothing he is left all alone. Since he does not want to be left alone, he will stop the grabbing behavior.

Repeat this exercise several times a day over the next week. By the end of the week, he will understand that grabbing clothes or shoes is not an acceptable behavior. Always mark and reward him with your attention when he stops himself from grabbing or when he simply does not try to grab your clothing. This is the behavior you wanted and you need to let him know that.

Treats

Grabbing treats is another area you may need to address with your puppy. Most people do not want their puppies biting them when they offer treats. If your puppy lunges or grabs treats from your hand, here is a way to stop this behavior. Take a dog bone treat and place it between your index and middle finger. Position the treat so the longest part of the dog bone is in the front of your hand. Offer your puppy the treat with the front of your hand facing the dog. Before you give the treat, say the word "Easy" in an authoritative voice and offer him the treat.

If he tries to grab the treat from your hand, put the treat away and just say the words "Too bad" and walk away. In about 10 to 15 minutes, offer him the treat again, holding it the same way. Tell him "Easy" and offer him the treat again. If he tries to grab or lunge at the treat, again say "Too bad" and put the treat away for an hour or so.

Repeat this exercise many times over the next week until he consistently takes the treat from your hand gently. Once he is reliably taking the treat from your hand gently, it is time to put the treat between your thumb and index finger and offer him the treat. Make sure to hold the treat far enough back so if he does lunge or try to grab the treat he does not grab your thumb at the same time or rip the treat out of your hand.

Using this new way of holding the treat, say "Easy" in an authoritative voice, and offer him the treat. If he grabs or lunges for the treat, put it away and walk away from him. This time wait 30 to 40 minutes and repeat the exercise. Repeat this exercise until he consistently takes the treat from your hand in a gentle fashion. In time, he will learn how to take treats gently from hands.

Grooming

Brushing and Combing

Brushing your puppy is an important bonding time for both of you. This is a wonderful time to build trust in the relationship. Check the brush and comb you are going to be using to make sure they do not scratch your puppy's skin. Some combs and brushes have sharp edges, and they can hurt. When this happens, your puppy will become concerned every time you try to groom him. If you are using a slicker brush, use one that has cushioning for the teeth and make sure that the actual teeth have been rounded off so they are not sharp. By checking tools before they touch your puppy, many grooming concerns can be eliminated.

If your puppy is afraid of the brush or comb, and he tries to bite you or the grooming tool, stop brushing or combing him immediately. Offer him the brush or comb to feel and taste. If he wants to bite it, let him. Give him a minute or so to check out this new object with which you are trying to touch him. When he looks away, that is a sign that he has finished checking out the new item and you can begin again.

Once the puppy has had an opportunity to explore the brush or comb, slowly start to brush him again. If he pulls away or cries, immediately stop again. Talk to your puppy and give him some pets and treats to take his mind off what just happened. Do something else with him for a while and try again later in the day.

The next time you try to brush or comb him, offer a few tiny treats as you use the back of the brush or comb to slide on his body. This will offer him a different kind of petting feeling without teeth. Talk softly, reassure him, and offer treats to reward this quiet behavior as you are using the back of the comb or brush.

After a few strokes, turn the brush or comb over and lightly do one stroke. If he accepts the stroke, be sure to praise him and keep the treats coming. If he tries to bite, cries, or squirms, stop again. Ask him for a simple cue such as *sit* and then mark and reward him and end the grooming session on a positive note. You can try again later the same day or the next day. For now, do not give the puppy too much attention and go about your normal daily activities.

Some puppies can be very sensitive about being groomed and will need time to work through this procedure. The most important piece of this exercise is to teach your puppy that you are paying attention to his concerns and stopping when he asks you to. In time, he will learn to trust you enough to allow brushing and combing. Remember that this is not a race, and it is important for you and your puppy to have a positive experience. Grooming your dog can be a wonderful bonding opportunity, and every dog needs grooming.

Work with your puppy using the brush and/or comb every day. Stop if your puppy cries, squirms, or tries to bite you. If the problem persists after several days, consult with your veterinarian because your puppy may have a medical issue you are not aware of that is causing this reaction. If your veterinarian does not find a medical reason for your puppy's reaction to being groomed, do not give up.

Give your puppy the time he needs to become comfortable with this procedure. In time, he will begin to trust you and learn that brushing or combing will not hurt him. When this happens, he will become comfortable with being groomed and this can become a special time for both of you.

Ear Cleaning

Checking your puppy's ears and keeping them clean can prevent infections and funguses. A number of products are on the market to keep your puppy's ears clean. Ask your veterinarian which product is best for your puppy.

Floppy ears can be a haven for yeast to grow, and this can be very uncomfortable for your puppy. To clean your puppy's ears, only use products specifically made for dog ears. Baby oil, rubbing alcohol, or non-ear cleaning solutions are *not* recommended. Products such as these can cause more harm than good to your puppy.

Many ear problems have a strong odor. If your puppy's ears have an odor, make an appointment with your veterinarian. Ear problems need attention as soon as possible. The longer you wait to take your puppy to the clinic, the worse the problem will become and the more painful it will be for him.

When checking your puppy's ears, the first thing you will want to do is to get him comfortable with having his ears touched. You can start off by petting his ears. If he does not like you touching his ears, then stop petting them for now. If he pulls away at any time while you are touching his ears, do not force him to accept the touch. This will only make him more concerned. Stop what you are doing and offer him a few little treats to take his mind off your touching his ears.

Offer your puppy a treat with one hand, and pet his ear with your other hand. If your puppy is now comfortable and does not pull away, you can lift up his ear and put it back down. If he still seems comfortable, you can look inside each ear and smell it.

If he is sensitive to his ears being touched, you will want to desensitize him to having them checked and/or cleaned. You can do this by spending a minute or so a day touching his ears. Do this several times on each ear until your puppy seems comfortable and shows no signs of concern. Next, try picking up one ear flap between your thumb and fingers. Lightly hold onto the ear and gently slide your hand out to the end of the ear. If the puppy shows any concern by crying, squirming, or trying to bite you, that is his way of asking you to stop. When you honor his concerns, over time, he will learn that you can be trusted. Once trust is established, life can be easier for both of you.

A few minutes later you can try again. Again, pet his ears with one hand while offering him a treat with the other hand. Try again to pick up the ear with one hand, and slide your hand down the ear gently while offering the treat with the other hand. Repeat this as many times as you need to, stopping every time he shows signs of concern. If he is still concerned after one or two minutes, stop for now, give him a nice pet along his back, or do something else you know he enjoys and try again later. You always want to end training sessions on a positive note.

Within a few days, the puppy will become more comfortable with having his ears touched. Once he is comfortable with the ear slides, pick up one ear at a time and look down into the ear canal. You will not be able to see all the way down into the canal, which is why smelling your puppy's ears is so important. The odor will alert you that something is wrong. Sometimes you will see redness or the ears seem to have a buildup much like ear wax in humans. This can be a sign of a fungus in the ear. If your puppy's ears have a bad odor or seem discolored, make an appointment for him to see your veterinarian immediately.

Checking your puppy's ears at least once a week and following an ear-cleaning regimen that your veterinarian recommends will help keep puppy's ears healthy and clean.

Nail Trimming

Many puppies have a concern with having their nails trimmed. This is actually very understandable. To your puppy, his feet are the major means of escape when in danger of any kind. With that in mind you, will want to let him know that touching his feet or trimming his nails is safe for him.

Your puppy will need nail trimming throughout his entire life. When nails are not properly trimmed, the living center of the nail (the quick) will continue to grow along with the nails. When this happens, it can cause discomfort and result in physical problems.

When trimming your puppy's nails, the first thing you will want to do is to get him used to having his feet touched by desensitizing him. The easiest way to accomplish this is by following your puppy's lead. Always move slower if you have a fearful puppy. Fearful puppies should be encouraged and rewarded for being brave with any new experience.

You can begin these exercises while you are holding your puppy in your lap or while sitting on the floor with him. Learning moments are everywhere for young puppies, so take advantage of them whenever possible with short learning moments.

Take your hand and slide it down your puppy's leg and pick up one paw at a time. If he is okay with you doing that, then take a finger and slide it between each toe of that paw. If he is comfortable with you playing with his feet and sliding your finger between each toe, you can repeat the exercise on each paw. If he pulls away or cries, stop immediately.

Wait a minute or so and offer him a treat with one hand as you slide the other hand down his leg again. If he pulls away, stop. If he is more interested in the treats than what you are doing with the other hand, then continue. Every time you start and stop, make sure you start sliding your hand from the top of his leg. This is an exercise of trust and confidence, so always start at the top, where he was comfortable. Then work your way down toward the paw.

Remember, if your puppy is not comfortable at any time, it is okay to stop and try again later that afternoon or the following day. If you stop when he shows concern, over time he will begin to trust you. Continue to repeat the exercise until all of his paws have been desensitized to your touch.

Once he is comfortable, the third or fourth time you touch his paws and slide your fingers between them, it is time to introduce the nail clippers. Let your puppy sniff, lick, or bite the newly introduced clippers. Once he is done exploring the clippers and ignores them, it is time to introduce the new sound the clippers will make. This can be done by opening them and closing them before you actually use them. Make sure you do not have any part of the puppy in the clippers while opening and closing them. If the puppy is bored and looks away, then you can begin the nail trimming process.

When trimming your puppy's nails, you want to clip them right before the quick. If he has clear nails, you can see the pink color of the blood vessel. If his nails are black, start near the tip of the nail and slowly clip small pieces at a time, moving up the nail. A tiny black circle in the center of the nail should be your warning that you are at the beginning of the quick, so stop. If you are not sure where the quick is or at what angle to hold the clippers, ask your Patient Behavior Advocate for some guidance.

If at any time your puppy pulls away, remember to simply stop what you are doing. Play with him for a while and try again later. You do not have to trim all of his nails on the same day when you are getting him comfortable with the nail trimming process. For now, the focus is on desensitizing him to having his paws touched and nails trimmed.

With time and patience, your puppy can become very comfortable with getting his nails trimmed. If you force him to have his nails cut, you will instill fear that can last a lifetime, making nail trimming a very difficult process for you and for him. Once mature, some dogs actually end up having to be sedated to have their nails trimmed. This makes a simple process very difficult and costly. Investing the time in your puppy now will eliminate this problem in the future.

When you trim your puppy's nails, always have a product such as Quik Stop® on hand in case you do cut the quick by accident. This product helps to stop the bleeding quickly.

Teeth Cleaning

Cleaning your puppy's teeth regularly will promote healthy gums and clean-smelling breath. Many puppies, though, are a bit concerned with having someone inside their mouths. Since your puppy's teeth will need daily attention, you will want to get him comfortable with you working on them.

To make this a pleasant experience, you will want to desensitize his mouth. Start off slowly and take your puppy's lead. Whenever he pulls away, squirms, bites, or cries, stop what you are doing. If you stop when he shows any sign of concern, he will soon begin to trust you.

When your puppy's concerns are ignored and you continue what you are doing, he can become fearful or show signs of aggression, which makes teeth cleaning harder on both you and your puppy. By acknowledging his concern and stopping what you are doing, you build trust with him.

You can begin desensitizing his mouth by rubbing your finger on the outside of his lips gently. If he is comfortable with this light touch, you can slip your index finger into the mouth and rub his teeth and gums gently with your finger. After a few days, if you have stopped every time he showed any sign of concern, he will become comfortable with these touches. Now it is time to introduce him to his toothbrush.

Let him smell, lick, or bite on this new object. When he becomes bored with the toothbrush and looks away or ignores the brush, it is time to introduce it to his mouth.

Begin by wetting the brush a little and using the dental cleanser you received from our office. Never use human-grade toothpaste; it is harmful to your puppy. Wet the brush to make it a bit more slippery so it does not stick to his lips.

Once the puppy is comfortable with having his teeth brushed, you can add a cue such as "Toothbrush time" or whatever word or words you would like to use to let him know it is time to get his teeth brushed.

Since dogs are capable of learning by example, let your puppy be in the bathroom with you while you brush your teeth. When you are done cleaning your teeth, tell your puppy it is his turn to brush his. Many puppies will wait patiently and enjoy having their teeth cleaned.

Remember to move slowly when desensitizing your puppy's mouth. Stop when he shows any sign of concern. Repeat the exercises a few times a day over many days. Your reward for the time and patience you give him now will last a lifetime.

Home Alone

Dogs are social animals. As a result, many puppies do not like to be left alone. Young dogs from 8 to 14 weeks of age are simply verbal during this period when isolated. If the behavior is ignored when presented, it will dissipate over time until your puppy outgrows this period. Socializing your puppy is an important aspect of building his self-confidence.

Using a Crate

If you are concerned about leaving your puppy in the crate, it may help to know young dogs need to sleep about 18 hours a day. Even when you are home with your puppy, it is a good idea to put him in the crate a few times a day so he gets the rest he needs and you can control his environment.

If your puppy has had a chance to become comfortable in the crate while you are home, this will also help him become more comfortable when you leave. When you put your puppy in the crate, use one special toy your puppy gets only when he is in there. Hard rubber toys that you can put treats into will give him something to work on while awake in the crate.

Before putting your puppy in the crate, always give him an opportunity to relieve himself. This way, you will know if he starts to whine, cry, or bark when you put him in the crate it is not because he needs to relieve himself. It is because he does not want to be left alone. Your puppy is verbally trying to tell you to open the door of the crate to let him out.

Opening the crate for a barking, crying, or whining puppy is a big mistake. Since puppies are constantly learning, you will be teaching him that every time he cries, you will open the crate and let him out. If your puppy has not been properly introduced to the crate or you have any questions about crate training, please let us know and we can give you some helpful information on crate training.

When you are feeling guilty about leaving your puppy alone or in the crate, your puppy will pick up on what you are feeling. Long good-byes before leaving the house will only add to the problem. Without realizing it, you could be instigating the concerned or stressed behavior.

Medication and Homeopathic Therapy

You may want to consider medication or homeopathic therapy for your puppy if he becomes concerned when left alone. Ask your veterinarian which product would work best for your puppy.

Using a T-shirt

Another way you can boost your puppy's confidence is with the help of a T-shirt. T-shirts give your puppy a better feel of his own body and help him to relax. The T-shirt should go over his head and fit snuggly on him. You want a T-shirt that goes all the way down to the end of his rib cage. If necessary, cut the sleeves on the T-shirt so they do not confine his front legs. This way, when the T-shirt is on, it will allow him to move around freely and will not be uncomfortable. Shirts that have spandex in them are great for this. Put the T-shirt on the puppy for 15 minutes the first time, and then you can gradually work up the length of time until he can wear his T-shirt all day. Once your puppy is comfortable with the shirt, start putting it on him 10 to 15 minutes before leaving the house. To further ensure his self-confidence with being left alone, start using the T-shirt along with a confidence course you can put together at home.

Desensitization

Another way to address your puppy's concerns is to desensitize him to you leaving the house. You can accomplish this by first breaking down what you actually do before you leave the house. Once you understand your pattern, you can begin to habituate him to each of the steps you take.

If your pattern is to put your shoes on before you walk out the door, then put your shoes on and stay in the house. At first, he may become very concerned about you putting your shoes on. However, when you stay in the house, he will realize you are not leaving, and he will settle down. After a little while, take your shoes off. A half hour later, put your shoes back on again, but stay in the house. Repeat this exercise a few times a day until your puppy ignores you when you put your shoes on.

If the next step in your pattern is to grab your keys, then begin the key-grabbing habituating just like you did the shoes. Repeat each pattern you offer before leaving the house separately until he becomes comfortable with every step. Once he is comfortable with the separate steps, it is time to walk out of the house.

Remember to not give any long good-byes. Walk out of the house for just a minute, then return inside. Once he is quiet, let him out of the crate. Repeat this exercise a few times a day, extending the time you leave him alone. In time, he will realize that when you leave, it does not mean you are leaving him forever. It just means you are leaving for a little while. Once he is comfortable knowing you will return, his anxiety should dissipate.

If you have tried all of the suggestions in this handout and your puppy is still distressed when left alone, please contact our office to discuss additional options.

Housetraining

The following are the keys to successfully housetraining your puppy:

- Manage your puppy's environment.
- Keep the puppy on a feeding schedule.
- Pick up any food the puppy does not consume after 15 minutes.
- Always reward the correct behavior (eliminating) when and where it happens.
- Always be consistent.

Basic Rules for Housetraining

Introduce a cue (word or words) to him when taking him to the designated elimination area, especially if the puppy is being trained to go outside.

Do not take the puppy for a walk to eliminate. Instead, take him to a designated place to eliminate and give him about six feet of leash to walk around while you are standing somewhat still. Once he has done his business, mark the elimination that has occurred in the proper area and reward him with a treat or take him for a walk as a reward. If you take the puppy for a walk to eliminate, the puppy can easily become distracted with all the different smells and sounds, and he may wait until he comes back inside the house to eliminate.

The other reason a walk is not recommended for elimination is because puppies quickly learn that once they eliminate the walk is over. They will learn to hold it as long as possible so the walk does not end. As the puppy's ability to hold it grows, walks will take longer and longer while waiting for the puppy to eliminate. There will be times when you do not have the time to continue the walk, you will come back inside the house, and he will eliminate on the floor.

Bring a treat with you when you take him outside to eliminate. Offering special treats just for proper elimination can make the training easier. The only time your puppy gets this really great treat is when he eliminates in the designated area. The second he is done with his business, mark the elimination with a word like "yes" or use a click from a clicker, then reward the behavior with the treat. After rewarding the initial elimination, stand still and wait if you think the puppy needs to eliminate again. Once he is finished eliminating the second or third time, mark and reward the proper elimination each time he eliminates. However, do not get too excited when marking the elimination behavior or you might distract him. A verbal "Yes, good boy" in a soft voice will suffice. Avoid giving him the treat inside when he returns from outside. The puppy will want to return inside too quickly to get his treat. He will relate the treat to coming back into the house and not the elimination he just did. Then, instead of completely finishing all his business outside, he will want to go back inside to get his treat.

Manage Your Puppy's Environment

You must constantly watch your puppy when he is not confined to a room, space, or crate. Accidents happen when you try to watch the puppy and cook, watch TV, do homework, or talk on the phone. Young puppies require constant supervision to understand what is expected of them with their new families and to learn what the rules are.

When you are distracted, you may miss your puppy's warning signals that tell you he is looking for a place to eliminate. Some puppies will sniff the ground, others will circle, some will raise their tails higher than normal, some will sit by the door leading outside, and others will walk quickly and suddenly squat. Every puppy has his or her own style and signals. It is your job to learn your puppy's signals.

When accidents occur—and they will—do not scold your puppy. This is very important! Scolding will cause many puppies to hide when relieving themselves so they do not get in trouble. This is why many new puppy parents end up finding surprises behind the couch or under tables. Elimination mistakes are usually the result of the puppy not being properly supervised. Paying close attention to him when he is not confined to his space will help prevent accidents from happening in the first place.

When an accident is in progress, make a short sound such as clapping your hands together to distract your puppy. Quickly scoop him up, if physically possible, and take him to the proper elimination area either outside or a wee-wee pad inside to finish his business.

Stay with him until he is finished, and remember to mark and reward him with a treat or walk for eliminating in the designated area. Clean the accident area with a product that will eliminate the odor completely. Do not use any products that contain ammonia, however. They only encourage future eliminations in the same place. Use products that are made specifically for this purpose.

Remember to put your puppy in his crate or confined area when you cannot manage his environment. Most puppies want to keep their sleeping area clean and will try to hold it as long as possible before eliminating there.

Puppies need to eliminate on a fairly regular schedule: when they first get up in the morning, after a nap, after play periods, 5 to 10 minutes after drinking, 5 to 20 minutes after eating, before they go into their crates, when they first come out of their crates, and before going to bed at night. During waking hours, puppies may need to eliminate every hour or so.

Small dogs can sometimes be a little more difficult to housetrain. They are very close to the ground, and you may not realize when your puppy is actually eliminating until it is too late. Keep a close eye on little ones to help them learn what you expect from them. Manage their environment carefully.

Teaching Your Puppy to Communicate

If you are taking your puppy outside to eliminate, it will be important to teach him how to tell you he needs to go outside in the future. You can begin working on this now by teaching your puppy to *speak* (bark) to let you know he needs to go outside.

Offer your puppy a special treat and tease him with the treat until he barks. The second the puppy barks, say the word "speak," then mark the behavior by using a word like "yes" or a click from a clicker. Reward him with the treat for barking. Repeat this exercise several times until he will *speak* on cue.

After he has learned to *speak* on cue, every time you take him outside, ask him if he wants to go outside, and give the cue "speak." Mark the behavior and reward him with a tiny treat, then take him outside to his designated elimination area. In time, he will learn to tell you he needs to go outside by barking.

As a general guide, you can confine an 8-week-old puppy for three hours, a 12-week-old puppy four hours, and a 16-week-old puppy for five hours before he will usually need to eliminate. If he does not get a chance to relieve himself within that time frame, you may end up with him soiling his area. Do not get upset with him if this occurs; he simply could not hold it any longer. This was the result of human error, not your puppy's mistake. If he is sleeping, you do not have to wake him up to go outside. Wait for him to wake up on his own before you take him to his designated elimination area. Remember to take the treats with you when you go outside so you will be ready to mark and reward him for eliminating in the proper area. Housetraining takes time, patience, and consistency.

Housetraining Troubleshooting

Suppose it has been more than a month since your puppy had an accident in the house. You think your job is complete and your puppy is now housetrained. Then, more housetraining accidents start to appear. What happened?

- The puppy may have a medical problem, such as an infection, and needs to be seen by his veterinarian.

- You forgot to teach him a cue that lets the puppy alert you to the fact he needs to go outside. If this happened, go back to the basics and introduce the cue (word or words). Cues such as "Outside," "Let's go outside," "Do you want to go outside?" are appropriate. You can use any word or words you choose, just be sure to use the same word or words consistently. Your puppy can learn to respond by getting excited, barking, or sitting.

- You forgot to teach your puppy to communicate with you when he needs to go outside. This can be accomplished by teaching him to *speak* (bark), *sit*, or even ring a bell that is hung on the door you use when taking him outside. You may also use a bell placed on the floor for your puppy to ring to let you know he needs to go out.

- You take your puppy for a walk, and he comes in the house to eliminate. Since puppies are constantly learning, the puppy now realizes that once he eliminates, the enjoyable walk comes to an end. As a result, he holds it as long as possible. You run out of time to keep walking him and come back inside. The walk has ended and the puppy forgot to eliminate while outside or did not want to because he didn't want the walk to be over. Either way, the puppy eliminates in the house. If this has happened, return to the basics and take the puppy to the designated elimination area. Stand there for a few minutes and wait until he eliminates. If he does, mark and reward him. If he does not eliminate in the elimination area, take him back inside and confine him to either his crate or a designated area. Wait 10 to 15 minutes and repeat the exercise. This must be continued until he finally eliminates outside. Mark the behavior (elimination) with "Yes" or a click from a clicker. Now take your puppy for a nice walk as the reward. You will need to repeat this exercise every time he needs to go out over the next several days until he understands that walks happen only after the elimination occurs.

- Your puppy does half of his eliminating outside and the other half of his eliminating inside. This can happen when treats are given to the puppy inside the house instead of outside where the elimination occurred. The puppy thinks he is being rewarded for coming into the house and, in turn, he hurries to get back into the house for his reward. Your puppy cannot relate the reward to the desired behavior when the behavior is performed at one location and the reward is given at another location. Rewarding your puppy in a different location only confuses him. To address this issue, take the treats outside with you and be ready to mark and reward him as soon as the elimination occurs. If you know he is not done, be patient. Stand there and wait for the next elimination. Once it has occurred, mark and reward immediately at the location of the elimination. After a few days, he will connect the wonderful treats with eliminating and will want to do as much eliminating as possible while outside to receive the rewards.

If you are still having problems with your puppy soiling in the house, please contact our office. Your puppy may be dealing with a health issue.

Hyperactive Puppies

Puppies are full of energy and require lots of exercise. Just like children, they need lots of opportunities and time for play. The timing of exercise, attention, and playtime should be your choice, not the puppy's.

When your puppy tries to get your attention by jumping, lunging, biting, licking, barking, or nipping at your clothes, it is important that you ignore him and walk away. Paying attention to your puppy when he is demanding your attention is not a good idea.

Keep your puppy close to you when working with this behavior. Put your puppy on a leash so you will have more control. You may find it easier at times to step on the leash when he is acting out. This will protect you or others and allow you to ignore him when necessary. When you step on the leash, allow your puppy enough room (about two to three feet from the clip to your foot) so that he does not feel pinned down. If he acts in a way that warrants stepping on the leash, look away and wait out his demanding and excitable behavior. It is important that your puppy not be able to hurt you by jumping up, grabbing your clothing, or nipping at your skin. The more opportunities you have to work with this behavior, the faster it will stop.

Here are some basic guidelines to help you teach your puppy self-control and good manners:

- Never allow your puppy to initiate play with you or other family members. You or the other family members should always be the ones to start play, not your puppy. However, if your puppy brings you a toy and you have the time to play with him, ask him to sit and then play with him as a reward.

- You should decide when playtime is over, not your puppy. Keep play sessions short (about five minutes) so that you can be the one to decide when playtime is over.

- Offer petting, scratching, and other forms of attention only to quiet puppies. Any rambunctious behaviors are ignored.

If the puppy jumps up, cross your arms over your chest to protect your face, arms, and fingers, turn your back on the puppy, and take a step forward. If he jumps up again, step on the leash to make sure he cannot hurt you, and take all your attention away from him.

Take all attention away from your puppy if he tries to demand your attention. This includes nudging your arm with his nose or barking at you for attention. Do not say a word. Look away, stand up, and walk away from him.

When your puppy is settled and quiet, walk over to him, softly pet him, and say "Yes, good, quiet." If he jumps up on you, walk away and try again when he is lying quietly. When he accepts the quiet petting, mark the quiet behavior with "Yes, good, quiet," and reward him with a soft, long pet. This will send a clear message to him that he will get your attention and rewards when he is quiet.

When it is time to feed the puppy, ask him for a *sit* before offering the meal. If he jumps up, do not feed him. Place the meal in the refrigerator or in a cabinet and walk away. Wait a minute or two for him to calm down and repeat the exercise. Once he remains seated for just a few seconds, quickly place the food dish on the floor and release him from the *sit* position. This will help teach him patience and self-control. If you are not successful after trying a couple of times, feed your puppy his normal meal. Set aside time to work on the *sit* cue with him before trying this exercise again. (If the puppy is not proficient with the *sit* cue, ask your Patient Behavior Advocate for the handout on teaching *sit*.)

A demanding puppy becomes a demanding dog, and demanding dogs can become aggressive dogs quickly. When you take all energy and attention away from this demanding and excitable behavior, it will diminish.

If your puppy is so out of control you cannot manage him, put him in his crate for a time-out. This gives him some time to calm down and may be necessary for only a few minutes. If you put him in his crate, do not open the door of the crate when he is barking. Opening the crate when your puppy is barking means he is making demands and training you.

Remember, everything should be done when *you* say, not when your puppy demands it. After consistently working with him for a few weeks, you should begin to notice a change in his behavior.

Jumping

Many puppies will jump up on their new family members for attention. In response, many people will reach down and pet their puppies without realizing the consequences. Rewarding the puppy with attention when he jumps on you ensures the behavior of jumping on people will continue. As your puppy grows in size and weight, jumping can get out of hand, be uncomfortable to live with, and be potentially dangerous.

Never pay attention to your puppy if he does not have all four paws on the floor. If your puppy is already jumping, take him into a small room of the house and re-create the scenario that caused him to jump on you. The second he starts to jump up on you, quickly walk out of the room and close the door. Wait 20 or 30 seconds, go back into the room, and greet him just like you did before. If he jumps on you again, repeat this exercise several times until he can keep all four paws on the floor when you walk into the room. This may take a few weeks, so after a few tries in each session, ask your puppy to *sit*, then mark and reward him to end your training session on a positive note. If you walk into the room and he does keep all four paws on the floor, mark and reward him immediately by petting him or giving him an ear scratch.

As the behavior starts to diminish, you can put your puppy on a leash and let him out of the room. If he jumps on you when he is outside of the room, cross your arms over your chest and look away from him. Once all four of his paws are on the floor, mark the behavior with the word "yes" and give an ear scratch as a reward. Crossing your arms over your chest serves two functions: It ensures your face is protected and it protects your arms and fingers from being bitten or scratched by your puppy while he is jumping.

If crossing your arms and ignoring him does not seem to work, it is time to use a leash. Put a four- to six-foot leash on him. Give him about three feet of leash and then step on the remainder. When the puppy tries to jump up, the leash will cause him to self-correct. The second all four paws are on the floor, mark and reward with an ear scratch. Once you feel the puppy understands that you do not want him to jump on you, introduce treats when he self-corrects his jumping behavior without the help of a leash.

When your puppy tries to jump up on you, you can take a step forward into his space. This will cause him to want to back up. To back up, he will need all four paws on the floor. The second all four paws hit the floor, mark and reward the four paws on the floor.

You can also try taking one giant step backward. Your puppy will not have your body to lean on and his front paws will have to go on the floor in order for him to keep his balance. The second all four paws are on the floor, mark and reward with an ear scratch.

If your puppy jumps on guests coming into your home, you can ask them to take a giant step backward or take a step forward toward your puppy. The second the puppy's feet hit the floor, ask your visitor to mark the behavior and then give the puppy a little ear scratch as a reward. Guests should also be asked not to pet your puppy unless all four paws are on the floor.

Grabbing your dog's paws and squeezing, or kneeing your dog in the chest, are hurtful methods and in most cases are not successful training methods. You can also end up with additional behavior problems. For example, these methods may cause your dog to not trust you touching his paws, making nail trimming a difficult task. Kneeing your dog in the chest can cause physical harm to your puppy, perhaps even a cracked rib.

If you enjoy your dog jumping on you but others do not, you can train him to jump up on cue. This behavior can be trained by giving the jumping behavior a cue (word) every time he jumps on you. You can use the word "up," if he jumps up on you without being invited, then use one of the methods described above to stop or ignore the uninvited jump.

A jumping puppy can be a danger, and accidents can occur. Jumping puppies can accidentally knock over an elderly person or a child, causing physical harm. A jumping puppy can scratch skin, rip clothing, and sometimes cause unintentional bites to the face.

Even if your dog loves people, he still needs to practice self-control and good manners. Greeting people with all four paws on the floor is a much safer, gentler, and more appropriate way for your puppy to behave.

Leash Pulling

When your puppy pulls on a leash, he is trying to get from point A to point B faster than you are. When your young dog pulls on the leash, stop walking and wait for him to turn and look back at you. The moment he turns his attention toward you, mark the behavior (paying attention to you) with a word like "yes" or click your clicker and have him come to you for the reward. Consistency is the key. You must stop any forward movement every time he pulls on the leash. Wait for him to pay attention to you and then mark and reward the correct behavior. You can also take a step or two backward or change direction if he is still pulling on the leash.

Another way to address leash pulling with your puppy is to tether him to your waist with a leash. A six-foot leash should be used when using the tethering method. Make sure your puppy is light enough to not pull you off balance. You do not want to get hurt in the process. Begin using the tether inside the house at first. Once he has adjusted to being within the leash distance from you, try the tether outside. This method can get him used to going only a few feet ahead of you.

When the puppy is where you want him, next to your side or walking on a loose leash, gently give him a few little pats on the head or mark with "yes" or a click from your clicker so he begins to understand what you want him to do: to walk on a loose leash. Every so often, you can give him a treat for being such a good boy.

These methods work with some puppies, but not all. Many puppies have lots of energy and can experience a bit of a challenge walking at our slower pace. For these dogs, some wonderful training aids can help with leash pulling. Two of the tools you can use are head collars and halters. When using any kind of head collar, you will want to first desensitize your dog to wearing it. Some dogs are concerned when you first try to fit them with head collars, and a halter may be a better option to start with.

Your Patient Behavior Advocate will decide which head collar or halter is best for your puppy, based on the challenges you are facing and the bone structure of your puppy. Head collars and halters are training tools and should be used as such. Once your puppy has stopped pulling on the leash, you can gradually go from the head collar or halter to a flat-buckled collar, based on whether your puppy is still trying to pull.

When taking your dog for a leisurely walk as a reward for relieving himself, do not expect him to walk at a heel position. Heeling (dog taught to walk exactly at your knee and hold that position) is good to use in a crowded situation or in the obedience ring. It is not a normal position for any animal for walking. Heeling does not give the puppy an opportunity to check out all the interesting smells that attract his attention. It is believed that dogs learn many things about other animals through their ability to smell where other animals have been. Give your puppy that opportunity.

Leashes

To get your puppy comfortable with his leash, clip the leash onto the collar or harness. Once attached, take a few steps in front of him while holding the leash in your hand. Say the words "Let's go." If your puppy starts pulling away or freezes, offer him a treat a few inches in front of his nose. Keep the food just far enough away that he will have to take one step forward to get to it. The second the first step is taken, mark the forward movement with "Yes" or a click from your clicker and give him the treat for his bravery.

Repeat the exercise, but this time place the treat far enough away that he will need to take two or three steps to reach the wonderful treat you are offering. Coax and encourage him to move forward. Do not pull or force him to move forward with the leash.

Once he does take those steps, mark and reward the forward movement quickly. Repeat the exercise many times until your puppy is joyfully waiting for the next treat and willing to take more and more steps to reach it. Once the puppy realizes the leash is a good thing, then it is time for a real walk. Always have some treats ready to use as rewards in case he gets a little concerned along the way.

If you do not want him to pull you when on a leash, do not pull him. If he starts biting on the leash, ignore the leash biting and keep on walking. It is better to replace a leash or two now than many leashes later in his life. When your puppy receives attention from acting out when biting on the leash, then the behavior will increase in intensity and duration. If the leash biting is ignored, it will diminish in intensity and duration and it will stop.

Loud Noises

Loud noises, such as motorcycles, gunshots, thunder, loud music, garbage trucks, and so forth, can bother some puppies. If it does not bother you, nine out of ten times it will not bother your puppy. If it does bother you, then in many cases it will also bother your puppy. He will pick up emotional cues from you and act accordingly. Always show confidence in what you are doing with him. He will sense it and be more comfortable.

If your puppy is concerned about loud noises, introduce him slowly to different sounds. You can distract him with an ear scratch, a toy, or a treat. Gradually increase the intensity of the noise as he becomes comfortable at each level. Proceed at your puppy's pace; you want to desensitize him gradually to loud sounds.

If you live in a quiet place, you may want to expose him to a sound CD. Just because it is quiet today does not mean five years from now you won't be living somewhere with violent thunderstorms, backfiring trucks, or Fourth of July fireworks.

The process of desensitization can be used for any object or situation that frightens your puppy. Remember, it is a gradual process, and taking small steps is extremely important. If you have any additional questions, please contact our office, as desensitization and/or medication may be needed in some cases to help your puppy cope.

Pay Attention

It is essential for every puppy to learn his name. When your puppy learns to turn his attention to you when he hears his name, it can help with any behavior you are teaching him. Having his attention makes training easier.

To train him to pay attention to you whenever he hears you call his name, take a wonderful-smelling treat and let him sniff it. Once he smells that wonderful treat, say his name. At the same time, bring the treat up between your eyes. When he looks you in the eyes (where you have the treat), mark the behavior with a "yes" or a click from your clicker and give him that great treat. If you lose his attention, you may have moved your hand too quickly from his nose to your eyes or the treat you are using to lure him has no value to him. Find a treat he really enjoys when training this behavior. Try again, only this time move slower.

Again, the second he looks at the treat between your eyes, mark and reward him with the treat in your hand. Practice a number of times a day until he becomes accustomed to looking you in the eyes every time he hears his name and follows the treat up to your eyes.

Once he is consistently looking you in the eyes every time you show him the treat, it is time to begin losing the lure (the exposed treat in your hand). Try the same exercise without the treat in your hand but using the same hand motion. This time when he looks up at you, mark and reward him with a jackpot (many small treats given one at a time). Repeat this step several times a day over the next few days. When he looks at you every time you call his name using the hand gestures, it is time to move to the next step.

Now, with your hands at your sides, call his name. The second he turns to look at you, quickly mark the behavior and reward him with a jackpot of treats. Continue working on this behavior over the next few days. Once he is consistently looking you in the eyes every time you call his name, it is time to make it a bit more challenging for him.

While your puppy is lying on the floor playing with a toy, call his name. When he stops playing with the toy to look at you, mark and reward him with a jackpot of treats. Repeat this only a few times a day over the next few days. Once he is consistently willing to take his attention off his toy to look at you, it is time to add other distractions, but just once in a while. These other distractions could include calling his name while people are playing with him, while he is playing with other animals, or even while he is eating his meals.

Training your puppy to take his attention off anything he is doing to pay attention to you is a wonderful skill and something for you to be very proud of. When you have his attention every time you call his name, training new behaviors will be easier.

Shy, Timid, and Fearful Puppies

Fear is an unpleasant emotion for humans and animals. Keeping this in mind will assist you in being patient and understanding with your puppy as you help him build confidence through socialization and desensitization. It is extremely important that you move slowly in the socialization and desensitization processes. Always work at your puppy's pace. Try not to force, pull, or demand anything unless you, your puppy, or others are in danger. Never scold him harshly or punish him in any way. This will only make him more concerned with his new world.

Getting timid puppies to wear a collar, walk on a leash, or allow anyone to pet them can be very challenging. The best thing you can do is to let your puppy experience new things at his own pace. If he pulls away from you, stop what you are doing. If your puppy freezes, stop what you are doing. The faster you stop what you are doing, the easier it will be for both you and your puppy the next time the same situation occurs.

Dogs are very good at trusting us. We can build this trust if we let them know quickly we are listening to them (by watching and interpreting their body language) and we are willing to accommodate their fear by responding to them. Stopping what you are doing when your puppy gives you a signal that he doesn't want you to continue lets your puppy know you are paying attention to his concerns. Knowing you are listening will help him trust you and begin to trust others.

When Dealing with Strangers

Other people should not try to pet your puppy unless he first actively moves toward them for a pet. Many shy, timid, and fearful puppies are more willing to get a treat than to be petted by strangers. Have the new person offer your puppy a treat. If he does not want to take the treat from the stranger, have the person throw a treat gently toward him, so he will not have to get too close to the person to receive it. Do this many times with different people until the puppy is comfortable walking up to a stranger for a treat. Strangers can gradually drop the treats closer and closer to themselves until he feels comfortable getting the treat right out of the person's hand. Repeating this exercise a few times a day over a few weeks can help build your puppy's confidence.

The Do-Not-Carry Rule

Carrying your puppy from point A to point B should be done as little as possible. When you or other family members are constantly carrying your puppy, you are sending him a message that he cannot handle anything on his own. Although it may be difficult for you to not carry your puppy a lot, letting him experience life with all four paws on the floor will be better for him in the long run.

Sit

There are three basic steps when teaching your puppy this cue:

- *Lure.* Put a treat just above the puppy's nose and move the treat slowly up over his head so he will need to look up to follow the treat. When most puppies' noses go up, their rears go down into the *sit* position, which is exactly what you want.
- *Mark.* The second your puppy's rear hits the floor, say a word such as "yes" or give a click from a clicker to mark the behavior (rear on floor). This lets your puppy know that was the behavior you wanted.
- *Reward.* Give your puppy a tasty treat, but make sure he receives it while his rear is still on the floor or you will be rewarding the wrong behavior. The correct position is rear on floor to receive his reward. A reward reinforces the behavior the puppy gave you, in this case, it is the *sit*.

Once your puppy is consistent at sitting promptly every time you give him the command and show him the lure, change the lure into a reward. This can be accomplished by moving your hands the same way, but without any treats in them. Add the verbal cue "sit" at the same time you are moving your hand over his head. The second his rear hits the floor, mark and reward him quickly. With a little time and practice, you can stop using your hands entirely and merely give the cue "sit."

It is very important to change the lure at the beginning into a reward only for completing the behavior, or your puppy will listen to you only when he sees a treat. The first time he sits without you showing him a treat, jackpot him for a good job. Instead of just one treat, a jackpot will be four or five tiny treats in a row. This will let him know he did a great job.

The finished behavior has three basic steps: cue, mark, and reward.

- *Cue.* Say the cue "sit."
- *Mark.* The moment the puppy's rear hits the floor, offer a "yes" or a click.
- *Reward.* Give the puppy a treat to reinforce the *sit* behavior he just gave you.

Helping Your Puppy Generalize the Cue

If you are teaching your puppy to sit in front of you, start asking for the *sit* at your side. You are asking for a new behavior, as dogs do not generalize well. When you teach your puppy to sit at your side, train the same way you did when asking for the *sit* when he was in front of you. Once he is consistent in front of you and at your side, it is time to introduce new environments, such as a different room, then outside, then from across the room, and so forth. Each step takes time. You can also begin to add other places, being given the cue by different people, and adding distractions. Now he will understand that *sit* means put rear on floor.

Many new puppy parents think their puppies are stubborn, hardheaded, or have selective hearing because they will not *sit* when they ask for the cue. In many cases, however, it is because the puppy parent did not teach the puppy the cue in many places or with distractions. Another reason that some puppies do not *sit* on cue is because they were habituated to the cue *sit*. This can happen when puppy parents repeat the cue more than once by saying "sit, sit, sit." Repeating a cue habituates the puppy to the cue. If you must repeat the cue, take a step to the right or left first, get your puppy's attention, and then repeat the cue once.

Troubleshooting *Sit*

Some puppies resist the above method by backing up. You can address this resistance in several ways. You can ask for the cue in a corner so that the puppy is restricted as to how far he can back up, or you can try asking for the *sit* on a different surface.

For some puppies, you can use a rug or pillow as a different texture to sit on. This can make a difference when the cue is first being taught. Once the puppy is consistently sitting on the special surface, you can train him in new places on different surfaces.

If you find that your puppy seems to be resisting the cue *sit*, do not push or force his rear down to the floor. There may be a medical reason for this resistance. Instead, contact our office and set up an appointment for a complete examination.

Socialization

Socialization is the process of introducing your puppy to new people, places, things, and experiences he will likely encounter in his lifetime as part of your family. This may include exposure to the veterinary practice without needing to be examined; visiting nursing homes; going to parks; seeing children running, screaming, and playing; and hearing loud noises, such as trains, motorcycles, and gunshots. You will also want to give your puppy opportunities to meet children, babies, the elderly, and folks in wheelchairs; people using canes, big hats, sunglasses, and costumes; and people in uniforms, such as police officers, postal workers, and delivery truck drivers. Take your puppy for rides in the car, let him walk on different surfaces, go for a ride on an elevator, take a nice walk through a park, or go to a ball game. These are wonderful experiences for your puppy and can be great fun. Allow your puppy to become comfortable with one socialization opportunity at a time until he seems comfortable with each situation before moving on to new or different experiences. You will not want to overload him with too much information too quickly.

It is important that you introduce your puppy to other animal species (such as cats, rabbits, horses, or goats), as well as other dogs. Introduce him not just to other animal members of your family, or next-door neighbors, but to all types—big, small, young, and old. Before you introduce your puppy to other animals, make sure that the other animals are properly immunized. It is important the other animals do not have a problem with puppies, though, or you will defeat the purpose of this interaction.

When introducing your puppy to other people, never hold him to receive a pet. Instead, let him meet the person at his own pace. If he does not want to greet the person, do not force the experience. Thank the person for his or her time and move on.

Try introducing him to other people again and again until he is willing to go up to the person and receive a treat that you gave them to give to him. Once your puppy learns that other people are wonderful creatures, you have accomplished your socialization mission with people.

If your puppy is a smaller breed, make sure not to carry him everywhere. You are not protecting him; instead, you are telling him that he is too small to handle anything on his own. If you continue carrying him around, he may bark at other dogs, animals, and people for the rest of his life.

Pushing, pulling, or forcing your puppy in any way defeats the entire socialization experience. It is important that you build gradually on his successes. Socializing your puppy can be a wonderful and fun time for both of you.

A collar, a leash, car rides, sporting events, loud music, trains, planes, automobiles, stairs, and parties to go to are all new and exciting experiences for a puppy.

One of the best things you can do for your puppy is to enroll him in a puppy class if there is one available. Make sure the trainer does not use any harsh corrections on your puppy, and if the trainer tells you to do anything harsh to your puppy, leave the class and do not go back. These early months set the foundation for his future, and young dogs (under 12 months old) go through two to three fear periods. Emotional or physical harm done during the first year can last a lifetime.

Textures

You will want your puppy to be comfortable walking on, over, and through anything you would want to walk through. So introduce him to those textures while he is young. Some examples of textures you can use include grass, sand, cement, gravel, plastic bags, rocks, plastic bags with water sprayed on them (makes them slick), water puddles, bridges, collapsed cardboard boxes, ice, snow, and carpets. Let your puppy approach every new texture at his own pace to build his confidence.

Building Confidence by Using a Confidence Course

A small, easy-to-assemble confidence course can do wonders to build your puppy's confidence. The confidence course should consist of things he can walk on, over, or through. Be creative and use items already in your home. You can use a big plastic garbage bag and place it on the floor for him to walk on. You can use a mop or broom handle for him to walk over. You can use a hula hoop for him to walk over or through. Styrofoam blocks give your puppy something to step over. An umbrella can be used to help your puppy get over a fear of new objects. Be creative and use your imagination. As he becomes used to one new item, add a second item. Always introduce one obstacle at a time until he is comfortable walking on, over, or through the item before introducing him to a new item. You always want to move at your puppy's pace and build on his successes.

When using a confidence course, put his collar and leash on the puppy and ask him to slowly walk through the course. Many puppies, especially in the 6 to 18 months age range, want to fly through the obstacles; however, this does not help anything. When you take your puppy through the course, take a few steps and stop.

Pet him for a few seconds and take a few more steps. It is important that he does this slowly. You will want him to pay attention to what he is doing. Slow walking with frequent stops helps him to pay attention. Any item that offers a different experience will work, so use your imagination. In a few weeks, this can help many puppies be more confident, especially when left alone.

Socialization with Children

Puppies and children should never be left unsupervised. Although they often have an affinity for one another and form a very strong bond, it is still a good idea to keep an eye on them when they are together. Left unsupervised, a puppy may bite a child in self-defense. Without proper supervision, it is difficult to identify the instigator and correct the problem. Children are often unknowingly unkind to animals, and the puppy is wrongly blamed for his response to the unkindness.

To some puppies, children are noisy, fast-moving objects with tempting flying hot dogs for fingers. Some puppies take it all in stride, but others become overwhelmed with too much stimuli happening too quickly for their comfort level.

When introductions are made, it is important to supervise the introductions. Children must be taught how to interact with animals safely. At first, instruct children to wait until the puppy approaches them before petting. They should be taught to respect when the puppy pulls away from them and to never bother the puppy when he is in his crate.

For the initial introductions, ask your child to approach the puppy from the side, never straight toward the new puppy. Ask your child to stop about three feet away from the puppy and extend one hand out to the puppy with the palm down. Allow the puppy to come up to sniff the extended hand. Once your puppy stands next to your child, the child can begin to pet the puppy on his side. If the puppy backs away from the child, do not force the interaction. Giving your puppy the time he needs today will help build a strong relationship between your child and the new puppy. Proper introductions will ensure that your child and puppy develop a healthy bond and become friends for life!

When your child does get that opportunity to actually pet the new puppy, explain the importance of petting the puppy gently and speaking softly. During the early stages of developing a relationship between your child and the new puppy, it is important that the child be instructed to avoid petting the puppy on the head, as many puppies are head shy. Once the puppy becomes more comfortable with the child, pats on the head can be added if the puppy does not shy away from the hand reaching over his head. If the puppy pulls away, head pats should not be allowed for a bit longer. Over time, with proper supervision, your child and puppy will have a very special relationship. If you have a few children, introduce the puppy to one child at a time, not all at the same time. They will be very excited, but this is not a relationship you want spoiled. Time, patience, understanding, and consistency are the recipe for a wonderful relationship between your children and their new puppy.

If your puppy is shy, timid, or fearful you will need to move very slowly in building this bond. If the puppy pulls away from your child, explain to your child that the puppy is a little shy right now and will need time to be comfortable. You can let the child offer the puppy a treat. If the puppy walks up for the treat, that is a great start. If the puppy is afraid to approach the child, let the child drop the treat on the floor and take a few steps back so the puppy can get the treat. After a few treat opportunities, the puppy will become conditioned to the idea that when the child is near, good things happen.

If the treats do not encourage the puppy to go to the child, explain to the child that the puppy is not brave enough right now and the puppy may feel a little braver next time. Most children are very understanding about such timid behavior and are willing to wait.

As a safety precaution, tell your children they should never approach a strange dog without the dog owner's permission. Any contact with strange dogs should be supervised by you as well as the dog owner. The same approach outlined above should be made to strange dogs. Always approach a dog from the side and not head-on. Do not reach over a strange dog's head as this could be misinterpreted.

Stay

When teaching the cue *stay*, it is easiest to train from the *down* or *sit* position. Just like *down* and *sit*, the cue *stay* requires a release word. You can use the words "Free dog" or "Okay." If you are a fan of the movie *Babe*, you can even use the words "That'll do." Whichever word or words you choose to release your puppy from a *stay* position is fine as long as you remain consistent.

When teaching this cue, first ask your puppy to do *sit* or *down*. Say the word "Stay" verbally, and use the hand signal of an open hand, fingers together in front of his face. Stand up straight, be still, and wait a few seconds. Mark the behavior quietly with a "Yes" or a click from your clicker and then reward and release him from the cue.

Never offer the mark or reward for staying unless he has not moved. If you put him in a *sit/stay* position, the mark and reward are given only if he is still in the same position he started from. If he breaks the *sit/stay* before you get the chance to mark, reward, and release, do not offer the mark or reward. If he breaks the *sit/stay* after the mark but before the reward, do not offer the reward. When you offer the mark or reward when the puppy has moved before being released, you are rewarding the wrong behavior. This will only confuse him. You asked for a *sit/stay* so the *sit/stay* is what you mark, reward, and release.

When you begin training the *stay* behavior, you can gradually increase the length of time before you mark, reward, and release. Start off by asking for a three-second *stay*, and then ask for five seconds, then ten, twenty, and thirty seconds. Each time you ask for the behavior, remember to mark, reward, and release the *stay* cue.

It is important to keep eye contact with the puppy when training this behavior. Watch your puppy closely for signs that tell you he is going to break the cue. He may begin to wiggle or adjust his body, a sign that he may be losing interest. Once you see any of these signs, quickly mark, reward, and release him from the *sit/stay* (or *down/stay*). By paying close attention to his body language, you can help him be successful. If you do not see the signs, he may break the behavior before you have had a chance to mark, reward, and release him.

Once your puppy is staying for 30 seconds in one room of the house, begin to train the *stay* in other rooms until he has learned the cue *stay* in a few different rooms.

Now that he can reliably stay for 30 seconds, it is time to lengthen the space between where you are standing and the puppy. Start off asking for the *stay* right in front of him. Once he is in the *stay* position for a few seconds, repeat the word "Stay," as well as the hand signal; then take a step back away from him. If he is staying quietly after a few seconds, step back into your original position. Give the mark, reward, and release. If he breaks the *stay* cue when you take the step back, simply return to him and put him into the exact same position again. Ask for the *sit/stay* in the same location you did the first time so he can begin to understand that *stay* means to stay exactly where you put him.

Next time you train this behavior and you get ready to take a step away, make it a very small step. Repeat the *sit/stay* cue verbally and with the hand signal. If he can stay for just a few seconds, step back into your original position. Mark, reward, and release him. Always give your puppy a chance to build on his successes.

Repeat this exercise many times over many months while gradually extending the distance between you and your puppy. Each time you train the cue *sit/stay*, start off in front of him. As you begin to lengthen your distance, take steps backward. Do not turn your back on him yet. The cue *stay* must be well established before you can turn away from him when you ask for the cue. Right now, he is just learning to deal with the distance between the two of you.

When the puppy can hold the *sit/stay* and you can back away 10 or 20 feet, he is ready for the next step in training this cue, adding distractions.

While he is in a *sit/stay* or *down/stay* position and you are across the room from him, repeat the *stay* cue both verbally and with the hand signal. Take a few steps to the right, stop, and face him, repeating the hand and verbal cue. If you notice him twitching, he may be getting concerned and ready to break the cue. Repeat the cue "Stay" as you walk back to where you started. Encourage the *stay* every step of the way, and if necessary repeat the cue "Stay" until you are standing in front of him. Once you are in front of him, quickly mark, reward, and release him. Repeat this exercise many times over many weeks until you can take steps to the right and left and he holds the *sit/stay* or *down/stay* position.

Next, you can make it a little more challenging for the puppy. When he is in a *sit/stay* or a *down/stay*, take a ball and gently roll it across the floor. If you see him squirming and wanting to go get the ball, repeat the cue both verbally and with the hand signal as you walk back to him. Once in front of him, mark and reward with many small treats this time for a job well done, and release him. Once he has been released, let him go get the ball and play together for a few minutes. Repeat the exercise later that day or later during the week until he can be still when he sees the ball roll across the floor.

With all training exercises it is important to remember that dogs do not generalize well. As a result, they must be taught the same cue in many different locations, possibly with different people, and with many different distractions until they can understand *stay* means stay where you are, even if toys, children, and wonderful treats are just a few feet away.

The cue *stay* can be a real challenge for puppies to learn. (Humans, too, can have a challenge with learning to stay still.) Be patient, and take small steps forward in your training. It is your job to help your puppy be successful at learning our language. Whenever training, always end your training sessions on a positive note. If he is having problems staying still, ask him to *sit*, and mark and reward him for the *sit*. For now the training session is over. Short sessions that end on a positive note, sprinkled throughout the day, will achieve wonderful results.

Toy Guarding

Puppies should learn from the beginning of their lives with their new families that humans are the givers of all good things and, when necessary, the takers of all good things. If your puppy growls when he is being approached while playing with a toy, address the behavior now. Puppies do not grow out of bad habits—bad habits just get worse if not addressed. You must remind your puppy who the toy giver is if he feels the need to guard toys.

If your puppy guards all of his toys, then all of the toys must be taken away. If your puppy guards one specific toy, then that specific toy must be taken away. This can be done while your puppy is in the crate or in a secure area.

Once all the toys have been put away, offer your puppy his least favorite toy to play with for a minute or so. Then, using one of your puppy's favorite treats, give the cue *drop-it* or *leave-it*. Whichever cue you choose, stay with that cue; do not change to a new cue later on. When your puppy drops the toy for the treat, mark the behavior with "yes" or a click from your clicker and give him the treat in exchange for the toy. Repeat this exercise just a few times a day.

Once the puppy begins to quickly respond to the *drop-it* or *leave-it* cue, it is time for the next step. Put away the toy you were using, and the next day bring out another toy he likes to play with, but not his most favorite toy.

Repeat the exercise with the new toy just a few times a day over the next few days, until he quickly drops the toy when he hears the cue "Drop-it" or "Leave-it." Once he will consistently listen to your cue of "Drop-it" or "Leave-it," it is time for the next step.

Take your puppy's favorite toy out and give it to him. Give him five or ten minutes to enjoy his toy, then walk over to him and give the cue "Drop-it" or "Leave-it," and offer him a very special treat. If he does not drop his toy, you can use a higher-valued treat, such as chicken or cheese, to lure the toy away from him. Once he finishes the treat, give him back his favorite toy and walk away. For now, let him play with his toy undisturbed for at least a half hour. Then repeat this exercise a few more times that same day, always allowing him time to enjoy his toy for a while first. Repeat the cue with his favorite toy a few times a day over the next few days.

Remember to mark the release of the toy, reward him with the treat, and give him the toy back to play with. Over time, he will learn to quickly drop whatever is in his mouth when he hears the cue. Once he gets to this level, rewards should be intermittent: one reward this time, three rewards next time, zero rewards the next time, and two rewards the next. Keep him guessing. Will he get one, three, or no treats when he gives the proper behavior?

Unless your puppy has something dangerous, never pull the toy out of his mouth. This will only cause him to want to guard the toys more fiercely, and you could get hurt.

If you offer the cue "Drop-it" or "Leave-it" to your puppy and he growls at you at any time, just say "Too bad" and walk away. Do not look at, play with, scold, or say any more words to him. For now, leave the room and completely ignore him. If he follows you to another room in the house, completely ignore him for the next five minutes. When it is time for him to go outside or to eat a meal, pick the toy up and put it away. In fact, if he continues to growl at you after you have tried the above methods, and the only toy he still guards is this favorite toy, simply throw the toy away.

You will not want to punish him for growling, as a growl is an important early warning system dogs give that asks others to stay away. A growl is the prerequisite for the bite. If you are concerned for your safety, stop and walk away.

If at any time your puppy becomes aggressive and tries to bite you while shaping this behavior, contact our office. It is extremely important that this behavior is addressed now before it is too late.

Bathing Your Kitten

Giving your kitten a bath while she is young will help her feel more comfortable with baths in the future. Although cats usually groom themselves, sometimes—depending on how long their hair is—they will need a bath occasionally. Getting her used to baths now will make the experience easier for her in the future.

Be sure to use shampoos and conditioners specifically made for kittens. Be sure to protect your kitten's ears from getting water in them. You will also want to be careful not to get soap in her eyes.

Depending on where you choose to bathe her, be sure to introduce the water slowly. The kitchen sink works well with kittens and with adult cats if they are not too big. Turn the water on gently, opposite from the side of the sink where the kitten is sitting. This will allow her time to get used to the sound of the water.

While slowly running the water out of the faucet, put a handful of water over her back to give her a chance to experience the feel of the water on her coat. If she squirms or tries to run away, stop putting water on her but keep the water running on the other side of the sink. Do some touches on her face with your index or middle finger. Start on her forehead and do tiny circles, moving her skin around one and a quarter times in one spot. Work your way down to the top of her nose and just do a few slow circular touches until she calms down.

This will calm down most cats. Once she has calmed down, again try putting a handful of water on her back. Repeat this a few times until she becomes comfortable and is not struggling to get away. If after a few tries she is still fighting you, do a few more touches on her forehead, put her in a towel and gently rub her for a minute or so, and put her on the floor. For today, bath time is over. Practice this a few days in a row until she becomes comfortable with the water on her back.

It is very important that you, rather than the kitten, decide when she leaves. If you let her go when she wants to, she will train you to do things her way. This is an opportunity to let her know you are here to understand and help her, but events will happen when you say they will, and not when she demands.

If the kitten is comfortable with the handful of water on her back, then you can move the nozzle from the sink over to her and let it run on her coat gently. Let her get used to the feel of the water. Once she is willing to hold still for a few seconds, you can begin her bath.

For the first actual bath, do not use much soap. The first bath is more of an opportunity for her to get comfortable with being bathed. The following day you can give her a real bath, but do not do too much too quickly. You want her to get used to baths so she will not fight you in the future.

Counter and Table Surfing

Kittens are very curious animals. When something on a table looks interesting to them, they will want to go see what it is. These objects can include photo frames, candles, coasters, vases, or anything new to the kitten. For a kitten, all of these objects are possible play toys. Kittens like to jump up onto surfaces; this is part of their play drive, and an inherent instinct in kittens.

Most kittens enjoy keeping an eye on their territory from a high point in a room. One of the easiest ways to address this behavior is to use a cat tree or climber that allows your kitten to find a vantage point higher than the countertop so she can be above her territory. Putting a little catnip on the tree will entice her to climb the tree to the highest vantage point. If higher than your countertops, she will develop a preference for the tree. Place the cat tree in a location where she can see inside and outside if possible.

Some ways you can interrupt the counter and table surfing behavior are listed here:

- Make a loud noise, such as a clap or a snap.
- Use a spray water bottle. Spray the kitten when she jumps up. Make sure the kitten associates the spray with the act of jumping up and does not associate it with you. (Spray and then hide the bottle.)
- Remove any items your kitten is intrigued with from countertops or tables.
- Use heavier items that are not easily knocked over to decorate countertops.
- Consider using something to put under the item so it sticks to the countertop.
- Use a training device such as a Scraminal.®

Introducing Your Kitten to Your Home

When you first bring your kitten home, you need to take steps to make the adjustment easier on both you and your kitten. A new place is scary for your kitten, so remember to take each step slowly and work at your kitten's pace. Whether your home houses other animals or not, you should introduce your kitten to one room at first, then add another, and so on. The same holds true for new people and objects. Introducing a kitten to new environments and people is a process that should be done gradually.

When bringing your kitten into your home for the first time, she should be in a carrying case. Before you bring her into her room, you should kitten-proof it; do not leave small objects lying around. Cats are very curious animals, and if there are small objects around for the kitten to play with or chew on, in many cases she will. Anything that is in the room that dangles or hangs should also be adjusted so the kitten cannot reach it. In addition, set the room up with a litter box, food, water, and one or two toys to investigate. Bring her into the room while still in her carrier. For now, the door to the carrier should remain closed for about a half hour. At this point, you simply leave your kitten alone. Any family member who wants to stay with her can, but they must speak softly to the kitten and sit still.

In about a half hour, you should check on your kitten. If she is meowing or near the front of her carrier asking to be let out, you can do so in this one room. Make sure someone sits with her quietly and gives her the time and space she needs to be comfortable with her new surroundings. Open the door to the carrier and let her come out on her own. Keep an eye on her but do not interact with her unless she initiates it by coming over to you. Make sure you keep the door to the room shut; once she decides to come out of the carrier, she may enjoy exploring her new environment or try to hide somewhere you cannot reach her.

Over the next few days, leave the kitten in her new room and make frequent visits to play and pet her if she will allow it. Give her time to investigate you and come to you on her own. Make sure the room you leave her in has been kitten-proofed so she cannot get hurt or in trouble.

When your kitten begins to cry when you leave her alone, it is time to open the door of the room to which she has been confined. Make sure other doors throughout your home are closed. You will want to give her time to get used to different areas in your home slowly. Let her come out of the room on her own.

Introducing Your Kitten to Other Pets

Around the third or fourth day after you bring your kitten home and she is comfortable with the first room you gave her, put the kitten into her carrier and bring her into the family room or other room where your family spends time together. If you have more than one pet, introduce the other pets to the kitten while she is still in her carrier, one at a time. Let one pet out to sniff and look at the new family member. Stay with your pets and supervise their behavior. At first, most kittens will exhibit anxious behavior with the new animal. Give both animals time to look at and watch each other. If after a half hour or so the kitten is still in the back of her carrier, take her back into the original room, where she is already comfortable. If the other pet is a dog and making noises (for example, barking) toward the kitten, ask him to settle down and be quiet. When he does, mark and reward his quiet behavior.

Kittens can be somewhat skittish, especially when they have not had socialization opportunities. Loud noises, such as hearing a dog bark for the first time, can be frightening to your kitten. Dogs do bark at strangers, especially if they have not had socialization opportunities with cats prior to the new kitten. Your dog may get excited when meeting this new friend, and he could scare the kitten unintentionally. Do not scold your dog for barking; ask for quiet instead. If you would like information on how to train your dog to be quiet on cue, please let us know.

If your dog can settle down, then let the two animals look at and smell each other while the kitten is in her carrier until your dog gets bored with the new kitten and lies down or walks away.

If your kitten is being introduced to an older cat in your home, do the same thing. Let the resident cat check out the kitten, if she will, and make sure to give the older cat lots of attention. In many cases, the resident cat will snub the new kitten or hiss at her. This is because another feline has infiltrated her territory. Within a few weeks, this snubbing and hissing behavior generally dwindles and eventually ceases. In some cases, however, this behavior lessens but still continues. It may help to hold your older cat and give her lots of additional attention. This should help to stop the jealousy issues some cats have when a new kitten is brought into the home. Sometimes your older cat will never accept the new kitten. Your older cat will, however, learn to tolerate the new kitten, even if she does not like her. Give your resident cat lots of attention to reassure her she is still top cat in the house.

For the first encounter with the other pet(s), five to ten minutes is enough time together for introductions. Depending on how the animals react to one another, only one minute may be enough time. Watch the interaction, and if either pet becomes too aroused or concerned, put the kitten back in her room and close the door. You can try the introductions again later that day or the next. If the first introduction went well, repeat the exercise a few more times the first day.

The next step is to bring your kitten into the room with the other animal present and open the door to her carrier. Make sure there are places for your kitten to hide if she feels threatened. Again, allow only one resident pet to meet her at any given time. Supervise both of them closely. The kitten should be allowed to come out of hiding in her own time. This may take a while, so you will need some patience. Remember to not scold or reprimand your dog for barking. Being escorted out of the room and not being able to stay with you is reprimand enough.

Repeat this introduction exercise with all family pets, one at a time, until they begin to get comfortable with one another. In time, many will become great friends and even share mealtimes together. Once all the pets have had many opportunities to check out their new family member, it is time to allow more than one resident pet at a time with the kitten in the room.

At first, with two resident pets checking her out, the kitten will most likely be afraid and hide. If the other two pets are dogs, ask them to be quiet and settle down. If the dogs do not quiet down, take them out of the room. Repeat this exercise later in the day until the dogs can be quiet when they are with the kitten. When introducing your kitten to other resident cats, the same supervision is needed. Again, cats can be very territorial animals. Keep this in mind because your other cats may snub and/or terrorize your new kitten. If you can let them work this out on their own, it will be easier for both of them. However, if you feel one animal is in danger, your intervention may be necessary.

Depending on the animals' reactions to each other, the introductions can take a few hours or as long as a few weeks. This is the most important time for all the animals. They will all need time to adjust to each other, and that time should be given.

When introducing any other animal, follow the same set of guidelines. When raised together, many animals can learn to be friendly with one another. However, it is important to note that some animals may not get along. For example, if you have a pet bird, you should not leave your bird and your kitten together unsupervised. Cats are predatory animals and instinctively they will most likely try to eat your bird. However, sometimes even the most unusual relationships can be formed for a lifelong friendship.

Knocking Things Over

A kitten will sooner or later knock over and break something of value. In most cases, this is not an accident. Some kittens will actually stretch out their front paws to knock things over. It may seem like a strange thing to do; however, this is due to the kitten's curiosity and play drive. Kittens when left alone for extended periods of time will look for something to entertain themselves. One of the things they may do is knock things over. This is not done out of spite; it is the result of the kitten being bored. Kittens do enjoy company, and when left alone for extended periods of time, they simply get bored.

Here are some suggestions for addressing this behavior:

- Kitten-proof your home. Put items away so the kitten does not have anything to knock over.

- Offer the kitten a cat tree and place it by a window. Put the cat tree in a location that gives her an excellent high view of her territory. Rubbing a little catnip on the tree will pique her curiosity and she will investigate it. Soon she will begin marking it to make the tree her very own by rubbing on it, scratching it, and laying on it.

- Keep fragile items in a place the kitten cannot reach.

- If the kitten knocks items over only when left alone, take short trips out of the house. When you return, if she has not knocked items over, give her lots of attention and tell her what a good kitty she is. Gradually extend the length of time you leave her alone, ignoring her when she has knocked items over and giving her lots of attention and praise when she does not.

- Offer the kitten a different toy to play with and to entertain herself with when you are leaving the house. Toys filled with treats are great fun and stimulate kittens. Figuring out how to get the treats out of the toy is both entertaining and stimulating. When your kitten has toys that stimulate and entertain her, she has less reason to look for items in your home to knock over.

- When the kitten knocks items over when you are home, you can use a spray bottle and spray her, but only if you can catch her in the act of knocking items over. Or you can snub her and walk away. Do not give her any attention for at least 10 minutes. When using a spray bottle, it is important that the kitten is conditioned to the spray being bad and does not associate it with you, so hide the bottle the second after you spray her.

- If you have decorative items you want to leave out, consider getting some glue dots to stick under your decorative items to keep them in place even if she does try to knock them over.

- It is a good idea to consider heavier items for decoration purposes. The heavier the items are, the more difficult it will be for her to knock them over.

- Spend time interacting with your kitten. She needs and enjoys your attention.

- Provide play periods with family members; the kitten needs her locomotive and predatory play drive needs met.

Litter Box Troubleshooting

Kittens do not usually have to be trained to use a litter box once they know where it is and, of course, if it is in a convenient yet somewhat private location. However, there are some cases in which a kitten will not use the litter box. If the kitten is eliminating inappropriately, there is probably a preference problem. In these cases, some of the tips given below may help to determine the kitten's preferences. Wait a week or two between each change to allow the kitten to make a choice and show her preference.

You can try using a different litter. Some kittens prefer litters that clump because they are softer. When trying out a new litter, it is important that you get another litter box identical to the one you already have. Put the same amount of the current litter in one box and the new litter in the other box. Let the kitten make her own decision. Try this for two weeks before making a final decision on which litter your kitten prefers.

Try changing the amount of litter you put in the box. Some kittens prefer a lot of litter (four to six inches), and others prefer little to none. Try adjusting the level of litter over a week or two. Start off with just an inch or so of litter in one box and no litter in another. Over the next week, gradually increase the amount of litter you put in her box.

If you have more than one cat, get separate litter boxes for each plus one extra. Cats often do not like sharing their litter boxes.

You can also try to slowly change the box's location, little by little. It should be placed somewhere that is easily accessible for your kitten, offers escape routes, and is quiet and private because cats do enjoy their privacy.

Make sure to clean the litter box on a regular basis. Kittens do not like dirty litter boxes or dirty litter. Scoop the box out a few times a day and wash the litter box at least once a month. If urine or feces gets stuck to the box, the box should be washed immediately.

Check whether the litter box is large enough for your cat. Many commercial litter boxes are not big enough for many cats. If you think this could be a concern, consider a larger plastic container for the cat to eliminate in. A good general rule for litter box size is three times the length of your cat from nose to tail. Plastic storage boxes that are designed to go under a bed make excellent litter boxes. They are usually long enough for most cats and low enough for easy entry and exit.

Be sure the litter box offers an easy entrance and exit for your cat. If it does not, you may want to open up the side further or consider a new box.

When you first bring your kitten home, you should keep her in one room. This room should contain her litter box. As you gradually introduce your kitten to more areas of your house, the litter box can be moved with her. When she has access to the full house, you can move the litter box slowly, a foot or so every day until it has reached its new location.

If you have a large home, consider getting multiple litter boxes. Again, place these in areas where they are easily accessible and provide privacy and escape routes.

You can also begin to reward your kitten's good behavior! Pay close attention to your kitten. When she uses her litter box, give her a treat! This may help her want to use her box more frequently and stop the inappropriate elimination.

Any recent changes in lifestyle or home surroundings may cause your cat to suddenly stop using her litter box as well. Cats are very sensitive to any sort of change, and this may be the root of the problem. Sometimes simply moving furniture around is enough of a disturbance for a kitten to stop using her litter box. If you have moved furniture around, make sure your kitten knows where her litter box is and give her lots of assurance that everything is okay. This is, in fact, a great time to introduce a new litter box and reward her when she does use her new litter box.

Playing Safely with Your Kitten

Learning to play safely with your kitten entails paying close attention to what you are doing as a kitten parent. Cats have a tendency to be full of energy one minute and napping the next. Think of the term "catnap." We use this to describe a short and quick nap. The reason it is called catnap is because cats do this repeatedly throughout the day.

When your new kitten has overcome her initial fear of being in her new home, it is important for you to understand that she will start displaying her curiosity through energetic exploring. To help her display this energy in a positive way, you need to play with her. It is your job to help her get her mental and physical exercise needs met.

Kittens basically have two modes of play: predatory and locomotive. Predatory play includes behaviors such as pouncing, grabbing, chasing, and throwing things in the air. Locomotive play includes behaviors such as running, climbing, leaping, and finding places the kitten can go into and come out of quickly, such as a paper bag or a box.

To play safely with your kitten, give her toys that stimulate both modes of play. For predatory play, balls, fake mice (especially those that make noise), and laser pointers are excellent suggestions. Keep in mind, however, that you do not want to get any objects small enough that your kitten can swallow them. Objects with feathers, although fun for your cat, will end up in pieces around your house. Kittens enjoy tinsel toys too, but they should be offered only during supervised playtime. For locomotive play, one suggestion is to get a cat climber. These come in many different sizes and types. They can be used not only as a place to climb and run through but also as a scratching post or a place to sleep. You can also have a cat tree for your kitten to climb on, scratch, and lay down on so she can oversee her territory.

Social play is great for your kitten, too. However, never use your feet or hands as play objects with a cat. Your kitten cannot tell the difference between your appendages and his toy mouse if both are being presented as toys!

Social play is interaction between your kitten and people or other animals within the household. This type of play is important not only for getting along with the family members, but also if you want your kitten to be comfortable around guests in your house.

Pouncing On Legs

Some kittens like to pounce on legs or toes when people walk by them. This is part of kittens' normal predatory play drive, but can still be very annoying. Some kittens will do this to let you know they want to be played with. Either way, you do not want to reward this behavior with your attention. Yelling at her or caressing her is still attention and, to a large extent, both are exactly what she wants—your attention. This is one way the kitten can tell you she needs some interactive playtime with you or with some toys. Kittens love to pounce on things; it is their predatory play drive in action. Toys such as a laser light she can chase around the room and then pounce on when it is hovering in one spot allows her to complete her predatory play drive. If she does not receive this kind of playtime, you are leaving it up to her to find something to pounce on, such as your legs. Contrary to popular belief, kittens do need interactive time with their families. They do not enjoy being left alone all the time. They need love, socialization, and playtime with all family members.

Here are some ways you can address the pouncing behavior:

- Make a loud noise to startle her (e.g., clap, snap, shake a plastic bottle with coins in it, or make a loud sound such as saying the word "Ouch" in a loud voice) and walk away from the kitten.

- Catch the kitten letting you walk by her and not pouncing on you, and mark and reward her gentle behavior with a nice long pet, treats, or your attention.

- When you can see that the kitten is getting ready to pounce on you, quickly get up and walk out of the room while completely ignoring her. Later, when she allows you to walk by without pouncing, give her lots of love and special attention. This is a perfect time to play with her, using a toy she can pounce on to get her predatory needs met while she is being a good girl.

- Use treats, circular touches, or simply your attention as the reward for good behavior.

- Enrich the kitten's environment by offering a more interesting variety of toys to stimulate her mentally. If she is bored, she might think your legs are the only toys available to her.

- Exchange the kitten's toys frequently. Offer one toy at a time for a few days and then put that toy away and offer her another toy. When you exchange her toys, they will stay interesting to her. If toys are all left out all the time, she will become bored with all of them.

In many cases, when kittens offer inappropriate behaviors, it is because they are bored.

Scratching and Biting

Scratching and biting are a part of how a cat not only defends herself, but how she plays. If you use your hands to play with your kitten, she will scratch and bite them in time. She will not understand that scratching and biting you is an inappropriate behavior; after all, that is how she naturally plays with other kittens. If your kitten ever scratches or bites you, all attention and play must stop. The kitten will eventually learn that when she scratches or bites, she receives no attention and any playtime is over.

Kittens and adult cats know when their claws are out or in, so scratching you is never an accident. When the kitten scratches anyone, all attention stops and she must be ignored. Your kitten is also aware when she bites; this, too, is not an accident. Some cats bite as a way of showing affection; however, this is still not good behavior. Whether she bites hard or soft, all attention stops and you should walk away.

Among the ways you can address this behavior are the following suggestions:

- When you are playing with the kitten and she bites, the play stops immediately and you walk away from her. It is important to be consistent. Do not one time disallow the biting, and the next time allow it because she bit gently. Only you can make the decision on whether you want the biting to stop or continue. If you are grooming her and she bites you, you may have hurt her unintentionally and the bite was her way of saying stop. Place your thumb under her muzzle and place the rest of your fingers on her head with one hand. With the other hand just brush her once or twice gently and let her go. If you just stop grooming her the minute she bites you, you are sending her a clear message that you will stop whatever you are doing if she bites you. When you control her head for just five seconds and do one or two soft brushes and then stop and let her go, in time she will learn biting does not work. If you believe your kitten is being very aggressive with you, then share that with your Patient Behavior Advocate so appropriate guidance can be given.

- When your kitten scratches you, all attention to her must stop. If you are playing with her and she scratches you, it was not an accident. Kittens as well as adult cats know when their claws are in or out. If you were scratched, it was on purpose. You may want to get a few different scratching posts for her. Kittens have preferences as to what they scratch on, and cats throughout their lives will need to scratch. Have a few different scratching posts scattered throughout your home. When she does scratch you, it is important that you ignore her and walk away. Over time, she will stop scratching you and limit her scratching to her scratching pads or posts.

- When the kitten scratches you, you can make a loud sound, such as clapping your hands together or saying "Ouch" in a loud voice, to interrupt her, and then walk away. This will startle her, and cats do not like to be startled. Ignore her for a few minutes to let her know you do not like when she scratches you.

- Never use your hands or feet to play with your kitten. If you do, you are offering her your skin to bite and scratch, as a play toy. Only use safe cat toys to play with your kitten.

APPENDIX F:
SAMPLE CLIENT HANDOUTS
(ALTERNATIVE METHODS)

Touch Work

When doing touches, always breathe and be mindful of what you are doing. Use fast circular movement with a rambunctious or high-strung animal at first, and then gradually slow down your circles. You must meet an energy where it is to take it to where you want it to go.

Use slow touches on shy, timid, and fearful dogs. You may even want to use the back of your hand to do the touches. Animals seem to instinctively know the difference between the front and back of a human hand and are less threatened by the back of the hand. Perhaps they know at some level we cannot grab them with the back of our hands.

Slow touches should be done in a clockwise circular motion. Do no more than one and a quarter circles in any place at a time. When you do the touches, do not simply rub over the animal's coat. Instead, take the skin with you while doing the circular movements. Touches can be done randomly on the animal; a pattern is not necessary. Random touches that are easy to do and comfortable are usually what will work best.

Normal touches take two seconds—counting 1001, 1002. Fast touches on hyper animals are only about a half second. If an animal does not seem to like what you are doing, change the speed, the pressure, or the location; be creative, be flexible, and experiment. Pressure on the touches should be no more than the amount of pressure you would use on your own eyelids when your eyes are closed.

When doing touches, pause frequently (every three to five minutes) to allow the animal time to process what it just experienced. The intent of touches is to give warmth, awareness, balance, self-confidence, self-awareness, self-esteem, and self-control to animals, and it does. So you can understand why you would want to stop frequently and give the animal a chance to absorb what it is feeling.

Touches will work on kittens and cats as well as it does on dogs, horses, and just about any animal you can imagine. When working with long-haired animals, one method that works nicely is to hold a small amount of hair and make a circle while holding on to the hair near the skin or scalp. Then slowly release the hair by sliding it through your fingers and continue a long, slow stroke even after you are no longer holding onto the hair. Cats especially enjoy long, slow movements.